Ancestry Reimagined

Ancestry Reimagined

Dismantling the Myth of Genetic Ethnicities

KOSTAS KAMPOURAKIS

OXFORD
UNIVERSITY PRESS

Oxford University Press is a department of the University of Oxford. It furthers
the University's objective of excellence in research, scholarship, and education
by publishing worldwide. Oxford is a registered trade mark of Oxford University
Press in the UK and certain other countries.

Published in the United States of America by Oxford University Press
198 Madison Avenue, New York, NY 10016, United States of America.

© Oxford University Press 2023

All rights reserved. No part of this publication may be reproduced, stored in
a retrieval system, or transmitted, in any form or by any means, without the
prior permission in writing of Oxford University Press, or as expressly permitted
by law, by license, or under terms agreed with the appropriate reproduction
rights organization. Inquiries concerning reproduction outside the scope of the
above should be sent to the Rights Department, Oxford University Press, at the
address above.

You must not circulate this work in any other form
and you must impose this same condition on any acquirer.

Library of Congress Cataloging-in-Publication Data
Names: Kampourakis, Kostas, author.
Title: Ancestry reimagined : dismantling the myth of genetic ethnicities /
Kostas Kampourakis.
Description: New York : Oxford University Press, [2023] |
Includes bibliographical references and index.
Identifiers: LCCN 2022040733 (print) | LCCN 2022040734 (ebook) |
ISBN 9780197656341 (Hardback) | ISBN 9780197656365 (epub) |
ISBN 9780197656372
Subjects: LCSH: Ethnicity. | Race—Social aspects. | DNA—Analysis.
Classification: LCC GN495.6 .K364 2023. (print) | LCC GN495.6 (ebook) |
DDC 305—dc23/eng/20220923
LC record available at https://lccn.loc.gov/2022040733
LC ebook record available at https://lccn.loc.gov/2022040734

DOI: 10.1093/oso/9780197656341.001.0001

Printed by Integrated Books International, United States of America

To the memory of my beloved grandmother Maria; who was the source of all genealogical information of my family on my father's side, clarifying relationships to the utmost level of detail; who was the kindest person you could ever meet; and who passed away at the age of 102 years old when I started writing this book.

Contents

Preface	ix
Acknowledgments	xiii

1. DNA Ancestry Testing: What It Is and What People Make of It	1
2. Essentializing Social Groups: Nations	24
3. From Race to Ethnicity in Ancestry Testing	49
4. Genealogical and Genetic Ancestry	72
5. Using DNA Ancestry Evidence to Retrace History	91
6. We Are All Africans, Ultimately	102
7. More Related Than Distinct	123
8. Social Constructs versus "Natural Order"	144
9. Separating DNA from Culture	173
10. Finding Meaning in Our Ancestry Testing	195
Conclusion	211

Index	215

Preface

"I don't know how that happened. How did that happen?"

That was the reaction of Karen when she read her DNA ancestry test results.

"I am 61% Southeast African, 38% Cameroon/Congo, and then 1% Benin and Togo."

Two weeks ago, when she first appeared on the show *The DNA Journey*, Karen was certain about her ancestry.

"Oh, I'm 100%. I'm pretty sure," she replied very confidently, when the interviewers asked her what percentage of her ancestry was all East Africa. But she was not equally confident when she was about to receive her results.

"How have the last few weeks been?" the interviewers asked her.

"Confusing . . . I'm just lost. I know who I am. I'm African, East African. Very sure of who I am. . . . Yeah, I don't know. I think . . . I just realized that I didn't really question about finding out." After a while, Karen began reading her DNA test results, which were different than her expectations, not being able to hold her tears.[1]

Jay was similarly surprised.

"5% . . . German. And did you just put Turkey in there to wind me up?"

"No!" the interviewer replied emphatically.

Jay continued: "Germany . . . Germany. I told you I wouldn't want to be German or Turkish, and I'm both of them, so yeah . . . It's just a bit of a shock really to find out how many different places I come from"

Jay's ethnicity estimate was 55% Ireland, 30% Great Britain, 5% France-Germany, 3% Spain-Portugal, 3% Italy-Greece, 3% Scandinavia, and 1% Turkey.

Like Karen, two weeks ago Jay was certain about his ethnicity: "I'm proud to be English. My family have served and we've defended this country, and we've been to war for this country, so I know I'm English. My parents and grandparents have told me I'm English, so I don't see there being any other option for me to be anything else other than English."[2]

[1] https://www.youtube.com/watch?v=2SB6ZaqEaLQ (accessed January 19, 2022).
[2] https://www.youtube.com/watch?v=g5o9DmUYCJA (accessed January 19, 2022).

X   PREFACE

But the DNA test results he received seemed to indicate otherwise.

Karen and Jay were among 67 people who were selected in 2016 to take a DNA test and embark on what was called *The DNA Journey*, organized by Ancestry.com, the major DNA ancestry testing company, in collaboration with the travel site Momondo.[3] This was advertised as a journey into who people are, based on their DNA:

> Think you know where you come from? Think you know what you're made of? Our sense of identity stems from our upbringing, our culture, our family values, and the stories passed down from generation to generation. But how much do you really know about who you are and where you come from, and how does that influence how you look at the rest of the world?

The implicit message here seems to be that our upbringing, our culture, our family values, and our stories are not sufficient in order to "really know" who we are. Something else is needed—and this is DNA. Contrary to what they initially believed, neither Karen was found to be 100% East African, nor Jay was found to be 100% English. Rather they had ancestries from several different regions. The main aim of the present book is to explain whether DNA ancestry testing can indeed reveal our ancestry.

There was also another, very important, message in the advertisement of *The DNA Journey*:

> What they discovered is that they have much more in common with other nationalities than they could have expected and that we are all members of one global family.[4]

That we share more than we think with one another and that we are all family was a commendable key message of that program. A secondary aim of the present book is to explain why this is the case.

The way we envision ourselves and our ancestors, and in turn the way we delineate the genealogical communities to which we believe that we belong, is described as the *genealogical imagination*. This is formed during our upbringing by the stories we are told by our parents and grandparents; by the

---

[3] https://www.momondo.com/discover/momondo-the-dna-journey-how-it-was-made (accessed January 19, 2022).

[4] All quotations come from https://www.ancestry.com/corporate/blog/the-dna-journey-powered-by-ancestrydna (accessed January 19, 2022).

traditions and cultural practices with which we become familiar and which we may end up adopting; and by the history lessons we are taught in school and elsewhere about where the communities we belong to come from. More specifically, being East African or being English, as Karen and Jay respectively thought of themselves, is about our ethnic identity and our sense of belonging to a category with particular cultural features, such as a language, traditions, and perhaps a religion and a homeland. Our ethnic identity can be a defining feature both of how we think of ourselves and of how others think of us. It may also determine the rights and the obligations that we have. This is why it matters a lot to people.

DNA ancestry testing has recently entered the scene as a new player that has the power to overthrow what we know about our ethnic identity from tradition, culture, and history. The DNA profiles of more than 39 million people who have taken DNA tests, like Karen and Jay, could be found in the databases of the major DNA testing companies as of March 7, 2022:

- 20 million in Ancestry (dna.ancestry.com)
- 12,200,000 in 23andMe (www.23andme.com)
- 5,600,000 in MyHeritage (www.myheritagedna.com)
- 1,400,000 in Family Tree DNA (www.familytreedna.com)[5]

These profiles probably correspond to a smaller number of different individuals, as many people have taken tests from more than one company, but the number is still huge. Some of these people are immersed in genealogical studies; others do it just for fun, or because someone gave them an ancestry test kit as a gift. In any case, it is likely that if they receive surprising results, as Karen and Jay did, their perception of their identity might change.

One important reason for taking a DNA ancestry test that these companies promote is the lack of, at least complete, knowledge about one's ancestry. This is often presented as a *genetic ignorance* that needs to be addressed.[6] "Discover where your family is from without even leaving your living room" is the message of Ancestry. "Dig deeper into your ancestry" is the message of 23andme. MyHeritage tells you: "Amaze yourself. Uncover your ethnic

---

[5] International Society of Genetic Genealogy Wiki, https://isogg.org/wiki/Autosomal_DNA_testi ng_comparison_chart (accessed March 22, 2022).

[6] Nash, C. (2012). Irish DNA: Making connections and making distinctions in Y-chromosome surname studies. In K. Schramm, D. Skinner, and R. Rottenburg (Eds.), *Identity politics and the new genetics: Re/creating categories of difference and belonging* (pp. 141–166). New York: Berghahn Press, p. 144.

xii PREFACE

origins and find new relatives with our simple DNA test." "Begin Your DNA Journey: Explore the world of DNA and learn more about your ancestry," FamilyTreeDNA suggests.[7] All these companies tell you that by simply spitting into a tube or swabbing the inside of your cheek, you can find out a lot— usually for a very affordable price—about your origins and your ancestors several generations ago through DNA. In particular, the tests are presented as providing information about ethnic groups or geographical regions. Most companies refer to *ethnic origins* or *ethnicity estimates* in their results (with the exception of 23andMe that refers to *ancestry composition* instead). This is a powerful message, because it promises revelations about ethnicity made by DNA that nothing else can make.

But what should Karen and Jay believe, eventually? What their parents and grandparents had told them about their ancestry? Or what the DNA ancestry testing results seemed to indicate about it? Can DNA indeed reveal who we are and where we come from? And do we all have as mixed ancestries as Karen and Jay had? These are some of the questions that we are going to explore throughout the present book. But answering these questions is neither simple nor straightforward. My role here is that of a cartographer that brings to your attention the current and most relevant knowledge from different fields, such as human evolution, human population genetics, and genetic genealogy, as well as anthropology, history, philosophy, political science, and psychology, in order to gradually construct a rich and coherent view of what kind of information DNA testing can and cannot provide about ethnic identity, and about ancestry more broadly.

If you are wondering whether DNA ancestry tests can reveal who you are and where you come from, the answers you will find in the present book will surprise you but also enlighten you.

---

[7] https://www.ancestry.com/dna/; https://www.23andme.com/dna-ancestry; https://www.myheritage.com/dna?d=1; https://www.familytreedna.com (accessed January 19, 2022).

Acknowledgments

Writing this book has been a fascinating experience. I am indebted to Nadina Persaud, editor at Oxford University Press, for supporting this book right from the start and for steering it in the right direction. If the present book is easy to read, it is thanks to her meticulous editing and her persistence to make it one. I am also grateful to Katie Pratt and Sujitha Logaganesan for being very helpful while working with me on this book until its publication, as well as to Leslie Anglin for nice copy-editing of its text.

While writing this book, I was fortunate to receive very valuable suggestions or advice from many people, to whom I express my deepest gratitude: Eduardo Amorim, Stefan Burmeister, Nathaniel Comfort, Rob DeSalle, Michael "Doc" Edge, Jonathan Hall, Patrick Geary, Henry Gee, Mark Gerstein, Alan Goodman, Joseph Graves, Sheldon Krimsky, Jon Marks, Iain Mathieson, Kevin Omland, Erik Peterson, Peter Pfaffelhuber, Pardis Sabeti, Jeannie Stiglic, Ian Tattersall, and Krishna Veeramah. I owe special thanks to my colleague at the University of Geneva, Estella "Sim" Poloni, for her diligent reading of the whole manuscript and a long discussion that helped clarify, or drop altogether, several points. I am also indebted to Alan Goodman for his suggestion to put all the technical information into boxes. Finally, thanks to an anonymous reviewer for the Press who provided some useful suggestions.

The motivation for writing this book came from discussions with my father, Giorgos, and his attempts to reconstruct our family tree. His questions about and conceptions of roots, genealogy, and heredity made me want to write this book for him and everyone else who might have the same questions. My father's mother, my grandmother Maria, who lived for more than a century, was most often the source of his genealogical information. She always explained all relationships in the finest detail, and she always had a nice word to say about everyone. Her family tree was indelibly marked with two untimely deaths: of her brother when he was 25 years old and of her younger son (my uncle) when he was 29. Because of her passion for genealogy, which came to an end when she passed away as I started writing this book, I dedicate it to her loving memory.

1

DNA Ancestry Testing

What It Is and What People Make of It

Ancestry, Race, Ethnicity, and Nationality

I begin this book with definitions, because the concepts related to ancestry are often confused in the public discourse. Sometimes this is done because people do not pay attention to what exactly these concepts are about, and so they use them interchangeably. For instance, when the majority of a people in a nation are also considered to be members of the same ethnic group, nationality and ethnicity can be easily confused. Other times, different concepts are intentionally used to refer to the same category. For instance, "White" is a category that is described as race in the US Census and as ethnicity in the UK census. Therefore, I have decided to define these concepts right from the start, in order to clarify their differences and also show how their improper use may lead to misunderstandings.

Let us begin with the concept of ancestry. Generally speaking, this concept refers to an individual's history and through that to individuals, places, and stories of the past. However, as evolutionary geneticist Joseph Graves and biological anthropologist Alan Goodman have pointed out, it is useful to distinguish among three types of ancestry:

- *Cultural ancestry*, which refers to the linguistic, social, cultural, and religious traditions that an individual feels connected to.
- *Geographic ancestry*, which refers to the place, or places, where an individual's ancestors came from.
- *Genetic ancestry*, which refers to the connections that an individual has to other people through their DNA.

As Graves and Goodman have noted, these three types of ancestry may or may not be related.[1] It is possible that all these three types overlap, but it is

---

[1] Graves, J. L., and Goodman, A. H. (2021). *Racism not race: Answers to frequently asked questions.* New York: Columbia University Press, pp. 179–180.

Ancestry Reimagined. Kostas Kampourakis, Oxford University Press. © Oxford University Press 2023.
DOI: 10.1093/oso/9780197656341.003.0001

2 ANCESTRY REIMAGINED

not necessary that they do. The reasons why this might or might not be the case are considered throughout the present book. To these, we need to add a fourth type of ancestry:

- *Genealogical ancestry*, which refers to all the identifiable ancestors in the family tree of an individual.

Therefore, it becomes immediately clear that people may use the word *ancestry* to refer to very different features of an individual: culture, geography, DNA, or family tree. People may have one or more of these concepts in mind when talking about ancestry, and so it is necessary to be clear what one is referring to. Given all this, I want to emphasize two points that are crucial for understanding the implications of DNA ancestry testing:

- Genealogical ancestry and genetic ancestry are not the same; in fact, the latter is a subset of the former (which I explain in detail in Chapter 4).
- The categories often used seem to allow inferences from genetic ancestry to cultural ancestry and/or geographical ancestry; but this can be misleading. (I explain why this is the case in Chapters 8 and 9.)

Having shown that ancestry is far from a simple concept, let us look at the other concepts most often associated with it.

In any discourse about ancestry, there are usually three distinct concepts that get into play: *race*, *nationality*, and *ethnicity*. All three concepts usually refer to communities that have a common culture. However, they should not be confused with one another because they also have some additional, distinctive features:

- *Race* makes explicit reference to common descent and visible biological characteristics, such as in skin color, which are considered as the primary markers of difference.
- *Nationality* makes explicit reference to a state, or a similar political form, having an official status that is designated by particular documents (identity card or passport).
- *Ethnicity* makes explicit reference to a common descent and cultural characteristics, such as language, as the primary markers of difference.[2]

---

[2] Based on Fenton, S. (2010). *Ethnicity* (2nd ed.). Cambridge: Polity, p. 22.

DNA ANCESTRY TESTING   3

I consider all of these communities in the present book, but my focus is on ethnicity, for three main reasons:

1. Ethnicity is, in practice, a major, although not universal, criterion for distinguishing humans into social groups. In 2008, sociologist and demographer Ann Morning found that 87 out of 138 countries used some form of ethnic classification to group their inhabitants. Despite the diversity of conceptualizations of ethnicity (e.g., as "race" or "nationality"), the terms "ethnicity" or "ethnic" were used substantially more than the second term, "nationality."[3]
2. Ethnicity is what DNA ancestry test results claim to reveal—usually without defining the concept itself at all. Companies often make references to "ethnicity estimates" or "ethnic origins," and in most cases, ethnicity is identified with a particular geographical region.
3. Ethnicity could be considered as more fundamental than race and nationality, because people of the same race or nationality may have different ethnicities. For instance, two White (race) US nationals (nationality) could be of Irish or Italian origin (ethnicity). I am not suggesting that ethnicity is more fundamental though, but only that it might be perceived as such.

However, race has long been the focus of the discourse about ancestry and DNA,[4] and so it is necessary to keep in mind a crucial distinction between two different conceptions of race:

---

[3] Morning, A. (2008). Ethnic classification in global perspective: A cross-national survey of the 2000 census round. *Population Research and Policy Review, 27*(2), 239–272.

[4] Roberts, D. (2011). *Fatal invention: How science, politics, and big business re-create race in the twenty-first century.* New York: New Press/ORIM; Wailoo, K., Nelson, A., and Lee, C. (Eds.). (2012). *Genetics and the unsettled past: The collision of DNA, race, and history.* Brunswick, NJ: Rutgers University Press; Bliss, C. (2012). *Race decoded: The genomic fight for social justice.* Stanford, CA: Stanford University Press; TallBear, K. (2013). *Native American DNA: Tribal belonging and the false promise of genetic science.* Minneapolis: University of Minnesota Press; Yudell, M. (2014). *Race unmasked: Biology and race in the 20th century.* New York: Columbia University Press; Nelson, A. (2016). *The social life of DNA: Race, reparations, and reconciliation after the genome.* Boston: Beacon Press; Marks, J. (2017). *Is science racist?* Cambridge: Polity; DeSalle, R., and Tattersall, I. (2018). *Troublesome science: The misuse of genetics and genomics in understanding race.* New York: Columbia University Press; Suzuki, K., and von Vacano, D. A. (Eds.). (2018). *Reconsidering race: Social science perspectives on racial categories in the age of genomics.* New York: Oxford University Press; Saini, A. (2019). *Superior: The return of race science.* London: 4th Estate; Rutherford, A. (2020). *How to argue with a racist: History, science, race and reality.* London: Weidenfeld and Nicolson; Graves, J. L., and Goodman, A. H. (2021). *Racism not race: Answers to frequently asked questions.* New York: Columbia University Press; Abel, S. (2021). *Permanent markers: Race, ancestry and the body after the genome.* Chapel Hill: The University of North Carolina Press.

4 ANCESTRY REIMAGINED

- *Social race*, which classifies humans into groups based on their appearance and assumed ancestry, and which has been used to establish social hierarchies. The resulting racial categories are unstable, as their definitions and names have changed through time, but also very real as they have impact on people's lives, for instance, in terms of wealth and health.
- *Biological race*, according to which races are real, natural, and due to fundamental differences in the biology of their members. It has been well established, and I also show in the present book, that the existence of biological races is not supported by science (see especially Chapter 7).[5]

In the present book, wherever I use the word *race* I always refer to the concept of social race. I add the adjective *biological* wherever I refer to the concept of biological race, for the purpose of clarity. Because races are real, I refer to them with a capital first letter, without quotations marks (e.g., Black or White). But these terms refer to the social race, not to the skin color of a person.

Having clarified the meaning of the concepts used, let us now consider what DNA ancestry testing is and what people actually make of its results.

Looking for Ancestry in DNA

DNA is often perceived as our essence: something within that somehow determines who we are and what we do (see Box 1.1). DNA has been described as the "book" of life; as a "blueprint"; as a "program"; and as a lot more. Therefore, it is assumed that by analyzing it, we can find out more about ourselves by way of understanding what is hidden deep inside us that could even determine who we are. Hence DNA testing. We test our DNA in order to figure out the hidden information therein. Roughly put, there are two kinds of DNA testing. One is about health and disease. The idea behind it is that by relating DNA and disease, we will be able to predict whether one will develop a disease, or at least how likely this will be. These tests are useful for many diseases, which are nevertheless quite rare. But they also have limitations.[6] The other type of DNA testing pertains to ancestry, and its potential, validity, and limitations are the focus of the present book.

---

[5] Graves, J. L., and Goodman, A. H. (2021). *Racism not race: Answers to frequently asked questions.* New York: Columbia University Press, p. 3.

[6] See Kampourakis, K. (2021). *Understanding genes.* Cambridge: Cambridge University Press.

Box 1.1 What Is DNA?

DNA stands for deoxyribonucleic acid, which is a very long molecule. We can think of DNA as having the form of a ladder, which consists of similar building blocks (called nucleotides) that differ only in that they contain one of four "bases": adenine (A), thymine (T), cytosine (C), and guanine (G). These nucleotides are connected one after the other, like train wagons, thus forming what we can think of as the rails of the ladder. There are also connections between the nucleotides of the opposite rails, thus forming what we can think of as the rungs of the ladder (see the embedded illustration). The exact sequence of As, Ts, Cs, and Gs, in segments of DNA called genes, contains the information for the production of proteins and other molecules, which are implicated in our development and physiology. This is why DNA is also described as our genome. Because a DNA molecule consists of pairs of bases (A-T and G-C), its length is usually measured in base pairs (bp). The units usually used are one million base pairs (1 megabase pairs or 1 Mb) and one thousand base pairs (1 kilo-base pairs or 1 Kb). Human DNA consists of 3.1 billion bp, and because we have two versions of each DNA molecule in each one of our cells, the total amount of DNA in each human cell is more than 6 billion bp, in the form of 46 DNA molecules.

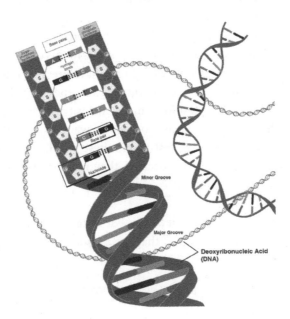

Courtesy of National Human Genome Research Institute (https://www.genome.gov/genetics-glossary/Deoxyribonucleic-Acid).

6 ANCESTRY REIMAGINED

In her exhaustive and detailed study of the companies offering direct-to-consumer genetic tests, legal scholar Andelka Phillips has concluded that approximately 30% of these companies offer tests related to ancestry. There exist companies offering other kinds of tests, such as those related to health, but the lines between the two kinds are becoming increasingly blurred—23andMe is one of the companies that does both.[7] This means that DNA ancestry testing is one part of a broad genetic testing enterprise, but of course a significant one as millions of people have taken such tests. But who are these people? Journalist Libby Copeland has usefully distinguished among three types of test-takers:

- *Answer seekers*: These are people who are trying to solve puzzles about their ancestry or about their biological relatives. This category includes, for example, people who have reasons to believe that the person they regard as their father is not their biological one, or people who are foundlings and who never knew their biological parents. These people are motivated to take a DNA ancestry test by their need to find answers to such personal questions.
- *Avid genealogists*: These are people who have been involved in genealogical research for a long time, and for whom taking a DNA ancestry test seemed a natural extension of their interest. These are actually people who may have taken tests from different companies and who have actively been searching for possible relatives on databases. Curiosity and interest are the most important motivations in this case.
- *Hobbyists*: These are people who do not have pressing questions to answer and who have no specific interest in genealogy. They may have taken a DNA ancestry test for fun, for curiosity just because a friend or relative also did it, or because someone gave the test to them as a gift. Among the three types of test-takers these are the ones who may experience the most surprise because they may be entirely unprepared for them.[8]

Some of these surprises are about family. For instance, novelist Dani Shapiro realized, after taking a DNA test, that the person she had regarded as her

---

[7] Phillips, A. M (2016). Only a click away—DTC genetics for ancestry, health, love . . . and more: A view of the business and regulatory landscape. *Applied & Translational Genomics, 8*, 16–22; Phillips, P. A. M. (2019). *Buying your self on the Internet: Wrap contracts and personal genomics.* Edinburgh: Edinburgh University Press.

[8] Copeland, L. (2020). *Long lost family: How DNA testing is upending who we are.* New York: Abrams Press, pp. 2–3.

father, and whom she had loved deeply, was not her biological father after all. It all started when her husband, Michael, decided to take an ancestry test as he wanted to learn more about his ancestry while his father was still mentally sharp, in contrast to his mother who had lost her memory. He thus ordered a test and also offered one to his wife. They spat in a tube, sealed it, and send it back to Ancestry, the company from which he had ordered the kit. But when Shapiro received her ethnicity estimate two months later, she was puzzled. According to Ancestry, her DNA was 52% Eastern European Ashkenazi (in short, Jewish). The rest was a compilation of European ancestries: French, English, Irish, and German. The reason that she was puzzled by these results was that she had been raised as an Orthodox Jew, like her parents and grandparents. Although she was not really religious, she was proud of that particular family history. But her ethnicity estimate based on DNA was not in agreement with that. She thus reached out to her older half-sister Susie, who was her father's daughter from an earlier marriage, and asked her whether she had taken a DNA ancestry test herself. When Shapiro and her husband compared her results to those of her half-sister, they were shocked to realize that they were not really sisters at all. Based on the comparison, it was estimated that their most recent common ancestor could be found 4.5 generations ago, not one generation ago as one would expect for any two siblings who have the same parents. This could only mean one thing: that the person they regarded as their father, who had died in a car accident many years ago, was the biological father of only one of them. Eventually, Shapiro managed to figure out that her biological father was a person (she did not disclose his identity in her book, but just called him Ben) who was a sperm donor at the time and place where she was conceived. She actually met him, and here is how she described her feelings: "We came from different worlds, Ben and I, and we had lived different lives—unshared lives—but everyone has a father, and he was mine. He begat me—to use the ancient language—and therefore a connection existed between us so powerful it felt impossible to grasp."[9] In this case, the DNA ancestry test had yielded an ethnicity estimate that made Shapiro reconsider what she took for granted throughout her whole life.

For others, revelations have little impact on their self-perceptions. Henry Louis Gates Jr., professor at Harvard University and director of the Hutchins Center for African and African American Research, had his whole DNA sequenced (see Box 1.2) along with that of his father—in fact, they were the

---

[9] Shapiro, D. (2019). *Inheritance: A memoir of genealogy, paternity, and love.* London: Daunt Books, pp. 3–11, 226.

Box 1.2 What Is DNA Sequencing?

In order to compare different DNA molecules, we have to figure out their sequence of As, Ts, Cs, and Gs (see Box 1.1) using methods described as DNA sequencing. These methods are often represented as being similar to "reading" a text that is already there waiting to be read, which makes them seem simple and straightforward. But they are extremely complex. The sequencing methods used nowadays are described as next generation sequencing (NGS). Here is very briefly how they work: DNA is first broken down into very small fragments, which undergo chemical reactions. While this happens, it is possible to identify which nucleotides are added whenever they are added, for instance either because the nucleotides carry a fluorescent molecule, or because a flash of light is generated. Thus, it becomes possible to eventually estimate the whole DNA sequence. The data outputs of these methods are valid and reliable, but errors are possible in DNA sequencing. To begin with, there exist some regions of the genome that are more difficult to sequence than others, and might thus not be represented in the final output. Other errors are due to the sequence reaction itself. One of these, called base calling, is due to the mistaken identification of nucleotides as the sequencing reaction proceeds. This can be due to different factors, such as the specific arrangement of the machine that does the sequencing. For such reasons, when it comes to DNA sequencing, it is important to consider sequencing coverage or depth: this has to do with how many times each individual nucleotide was independently sequenced. The more times this is done, the more likely it is to infer the "true" sequence. The take-home message from all this is that DNA sequencing is a lot more complicated than reading a text. DNA sequencing requires human intervention and manipulation of DNA molecules in order for their sequence to be "read." Can you imagine how it would be if you were required to modify and manipulate the text on this page in order to be able to read it?

For a very accessible account of all these technologies, see Rappoport, J. Z. (2020). *Mapping humanity: How modern genetics is changing criminal justice, personalized medicine, and our identities*. Dallas, TX: BenBella Books, Chapters 3 and 4.

first father-son pair, and the first African Americans, to do this. The analysis of his DNA revealed that Gates had "50 percent sub-Saharan African and 50 percent European and virtually no Native American ancestry."[10] However, this did not mean that Gates stopped considering himself as an African American. At the opposite side, in 2013 White supremacist Craig Cobb participated in the TV show's "Race in America" series, hosted by Trisha Goddard, having agreed to have his DNA tested. He had become famous for his (failed) efforts to purchase land in Leith, North Dakota, and establish a White supremacist enclave. When the hostess revealed the results that described him as 86% European and 14% sub-Saharan African, he immediately dismissed them as "statistical noise" and "short science," while also adding that "I will tell you this: oil and water don't mix."[11] He later reanalyzed his results to reject those that he received in the show.[12]

If you think about this, having a 14% sub-Saharan African ancestry is equivalent to having one great-grandparent who was sub-Saharan African. The reason for this is that each person shares 50% of their DNA with each of their parents. A parent shares 50% of their DNA with each of their own parents, which means that a person shares 50% × 50%, or 0.5 × 0.5 = 0.25 or 25% with each of their grandparents. Following the same logic, a person will share 50% × 50% × 50%, or 0.5 × 0.5 × 0.5 = 0.125 or 12.5% with each of their great-grandparents (see Chapter 4). There are numerous other such cases where people were surprised by their mixed results, because they might have thought that they had a pure ancestry—do not forget Karen and Jay whom we met in the Preface. So it is interesting to see if this is really what the people who take DNA ancestry tests believe and what results they get.

One study focused on the ancestry of 5,269 self-reported African Americans, 8,663 Latinos, and 148,789 European Americans living across the United States and drawn from the customer base of 23andMe, which as we saw is one of the largest consumer personal genetics companies. The main aim of the study was to clarify the relationship between the ancestry indicated by a DNA test and self-reported ancestry: Does everyone who reports themselves as being a member of a particular group found to indeed belong to that group based on their DNA? Participants were invited to fill in questionnaires, as well as to allow their DNA data and survey responses to

---

[10] https://www.npr.org/2019/01/21/686531998/historian-henry-louis-gates-jr-on-dna-testing-and-finding-his-own-roots (accessed January 19, 2022).

[11] https://www.dailymail.co.uk/news/article-2493491/White-supremacist-Craig-Cobbs-DNA-test-reveals-hes-14-African.html; https://www.youtube.com/watch?v=ptSZnTtGCQA (accessed January 19, 2022).

[12] https://www.stormfront.org/forum/t1092083/ (accessed January 19, 2022).

10 ANCESTRY REIMAGINED

be used for research. The authors of the study stated their awareness that "ancestry, ethnicity, identity, and race are complex labels that result both from visible traits, such as skin color, and from cultural, economic, geographical, and social factors." To get participants' self-perceptions of their ancestry, they asked two questions:

- *Ancestry/ethnicity*: "What best describes your ancestry/ethnicity?" from "African," "African American," "Central Asian," "Declined," "East Asian," "European," "Latino," "Mideast," "Multiple ancestries," "Native American," "Not sure," "Other," "Pacific Islander," "South Asian," and "Southeast Asian."
- *US Census categories*: "Which of these US Census categories describe your racial identity? Please check all that apply" from the following list of ethnicities: "White," "Black," "American Indian," "Asian," "Native Hawaiian," "Other," "Not sure," and "Other racial identity."

These two questions were used simultaneously in order to ensure the consistency of participants' responses. Overall, the participants were found to be consistent, and most self-reported a single ancestry/ethnicity category. Whereas the participants who identified as European Americans had on average more than 98% European ancestry, this was not the case for the other groups. Participants who self-identified as Latinos had on average about 65% European ancestry, 24% Native American ancestry, and 6% African ancestry. Finally, the participants who identified as African Americans had on average 73% African ancestry and 24% European ancestry.[13] I think that the most important conclusion from this study is that whereas most participants self-identified with one group, many of them were found to have ancestries from the other two groups as well. This is a key message that might suffice to debunk any notion of "pure" ancestry.

However, there are two important points to keep in mind. First, the constitution of the sample was 3.24% African Americans, 5.32% Latinos, and 91.44% European Americans. However, according to the US Census Bureau the population estimates for 2019 were 13.4% "Black or African American alone," 18.5% "Hispanic or Latino," and 76.3% "White alone."[14] This clearly

---

[13] Bryc, K., Durand, E. Y., Macpherson, J. M., Reich, D., and Mountain, J. L. (2015). The genetic ancestry of African Americans, Latinos, and European Americans across the United States. *The American Journal of Human Genetics*, 96(1), 37–53.

[14] https://www.census.gov/quickfacts/fact/table/US/PST045219 (accessed January 19, 2022).

shows the unequal representation of these groups in the 23andMe database. The people who are included in the reference groups of 23&me, and any other company, are those who have taken DNA ancestry tests, which in turn are those who first and foremost can afford these tests. Therefore, there is an important concern about the representativeness (an issue also considered in Chapter 8) of the reference groups on which the results of DNA ancestry tests are based (the related problems and implications are discussed in Chapter 10).

Second, the groups in the study just discussed are actually racial groups, as least according to the US Census. But what the DNA ancestry test usually provides are results framed in terms of ethnicity, for instance "Irish" and "Greek." It is therefore important to note that race and ethnicity can be confused in genetic ancestry testing. Usually, reference is made with respect to geography, that is, continental regions such as "Africa" or "Europe," and subcontinental regions, such as "East Africa" or "Southeast Europe." But the labels used to refer to these geographical locations are the labels also often used to refer to racial and ethnic groups. As reference groups become larger, the geographical resolution increases and reference is more and more made to subcontinental regions and ethnic subgroups. The problem therefore is that whereas the results are framed in geographical terms, which is sensible, if the description of the respective groups is made in terms of race and ethnicity (see Chapter 3), it can be confusing.

But perhaps the most intriguing problem that many people have had to deal with is that different DNA ancestry companies can provide different results for the same individual. Let us see why, and what this means.

Same DNA, Different Results?

In 2018, Charlsie Agro, host of the consumer investigative show *Marketplace* of the Canadian Broadcasting Corporation, and her sister Carly, a sports reporter, ordered DNA kits from AncestryDNA, MyHeritage, 23andMe, FamilyTreeDNA, and Living DNA in order to compare their ancestry breakdowns. They followed the required procedures and send their samples to each company for analysis. Charlsie and Carly are monozygotic twins; this means that they came from the same fertilized ovum that was later divided into two parts that each developed into a single embryo. As a result, they are supposed to have near identical DNA, with the exception of changes that might have occurred since their conception. Therefore, one should have expected their ethnicity estimates to have been exactly the same. But this was not the case, and furthermore each one's ethnicity estimate also differed from company to company (Table 1.1).[15] Let us then explore the emerging questions: Why did the ethnicity estimate of the same person vary from company to company (the columns in Table 1.1, each corresponding to Charlsie and Carly)? Why did two monozygotic twins receive different results from the same company (the rows in Table 1.1, each corresponding to one of the companies)?

The answer to the first question is relatively simple and straightforward. The twins received results that differed from company to company because of the different reference groups and different algorithms used by each company. If we compare the DNA of the same person to that of two different groups of people, we might get different results because one group could be more variable than the other. If we compare the DNA of the same person to the same group of people using two different algorithms, there would again be differences because of the different ways that the calculations are made with each algorithm. So you can imagine what happens when both the reference groups we compare a person's DNA to and the algorithms used for the calculations are different. There would exist two different variables that lead to different results for different reasons. Given this, the different results that Charlsie or Carly each got from the five companies should not be very surprising.

But the fact that two twins received different results from the same company is surprising. The mostly identical DNA sequences of the twins compared to the same reference group would be expected to yield similar results. Charlsie Argo visited computational biologist Mark Gerstein and his group at the Yale School of Medicine in order to understand why she and her twin sister received different ethnicity estimates. Gerstein replied: "I have to say, that one really shocked us. I mean, we expected two identical twins to have the exact same ancestry and they should. The fact that they present different results for you and your sister I find very mystifying. We thought for sure that the differences had to do when one person spit, there was a contaminant in the sample." However, when Gerstein and his colleagues looked at the DNA sequences of Charlsie and Carly, there was no difference. They

---

[15] Argo, C., and Denne, L. Twins get some "mystifying" results when they buy five DNA ancestry kits to the test. https://www.cbc.ca/news/technology/dna-ancestry-kits-twins-marketplace-1.4980976 (accessed November 8, 2021).

Table 1.1 The DNA Ancestry Results of Charlsie and Carly from the Five Companies

	Charlsie	Carly
AncestryDNA	39% Eastern Europe and Russia 27% Italy 23% Greece and Balkans 9% Baltic States 2% Turkey and Caucasus	38% Eastern Europe and Russia 29% Italy 23% Greece and Balkans 9% Baltic States 1% Turkey and Caucasus
23andMe	37.7% Italian 28% Eastern European 14.5% Greek and Balkan 2.6% French and German 1.5% Spanish and Portuguese 3.8% Broadly European 8.8% Broadly Southern European 1.1% Broadly Northwestern European 1.6% Western Asian and North Africa	36.8% Italian 24.7% Eastern European 13.5% Greek and Balkan 1.5% Spanish and Portuguese 12.7% Broadly European 8.3% Broadly Southern European 0.6% Broadly Northwestern European 1.8% Western Asian and North Africa
MyHeritageDNA	60.7% Balkan 19% Greek 13.1% North and West European 3.8% Middle East 3.4% Italian	60.6% Balkan 20% Greek 13.5% North and West European 3.1% Middle Eastern 2.8% Italian
FamilyTreeDNA	43% Southeast Europe 36% East Europe 6% Iberia 13% Middle Eastern	40% Southeast Europe 35% East Europe 8% Iberia 14% Middle Eastern
Living DNA	Northeast Europe 40.5% Tuscany 35.7% North Italy 4.3% Europe South 4.2% Europe East 2.7% Orkney Islands 2.3% Europe unassigned 7.3% Pashtun 2% Armenia and Cyprus 1.2%	Northeast Europe 47% South Italy 24.3% Northwestern 16.4% Europe 2.6% Europe East 1.7% Europe unassigned 1.8% Armenia and Cyprus 2.6% Pashtun 1.4%

All data extracted from https://www.youtube.com/watch?v=Isa5c1p6aC0 and reproduced here with the kind permission of CBC-TV.

14 ANCESTRY REIMAGINED

downloaded and compared the DNA sequences of Charlsie and Carly from all five companies, and found a 99.6%–99.7% agreement. As Gerstein put it: "No difference. It's shockingly similar. It's scarily the same." Therefore, the only explanation about the differences could be related to how the companies made their analyses. Gerstein concluded: "I think that the clean thing to say is we don't know how they did the calculations but we strongly think that you and your sister should get the same report, end of day."[16]

This brings us to an important issue I want to clarify: What we get from sequencing methods (Box 1.2) are DNA sequence data about individuals. These are data that can be used to make comparisons and eventually serve as evidence on which inferences about ancestry can be based (see Chapter 5 for specific historical examples). I have written that DNA "data" can be used as "evidence" because these terms are not synonymous. Data are just facts; they have no meaning on their own. When data are studied and interpreted, they can become evidence, that is, a reason to arrive at a particular conclusion. Data become evidence when we use them in favor of or against the justification of a belief. This entails that the same data can be used as evidence for or against different beliefs. Because scientists studying the same data can interpret it in different ways and draw different conclusions from it, DNA data cannot directly point to particular conclusions; rather, how it is interpreted is crucial. Evidence can support a conclusion, but it can never prove it. This is something important to keep in mind: DNA data "do not speak for themselves"; interpretations are crucial.

Perhaps inspired by the Agro twins' story, as they cite it in their article, a group of researchers at Case Western Reserve University analyzed DNA data from 42 monozygotic twins (21 pairs) who provided samples for three genetic testing companies: 23andMe, Ancestry, and MyHeritage. The aim was to estimate the "concordance of ancestry" (a) when twin pairs were tested by the same company and (b) when the same person was tested by different companies.[17] It should be noted here that because the different companies

---

[16] https://www.youtube.com/watch?v=Isa5c1p6aC0 (accessed January 19, 2022).

[17] The concordance for the twin pairs was calculated by using the lower of their two percentages for the same ancestry as a measure of percentage agreement. For example, if one of the twins was found to have 30% Italian ancestry and the other 40% Italian ancestry, the percentage agreement for Italian ancestry would be 30%. Similarly, if one of the twins was found to have 70% Scandinavian ancestry and the other 60% Scandinavian ancestry, then the percentage agreement for Scandinavian ancestry would be 60%. The total percentage agreement was determined by adding all these percentages for all ethnicities, in this case 30% Italian + 60% Scandinavian = 90%. A similar procedure was followed for comparing the results of the same participant across companies (36 of whom were female and all of whom were White).

do not use the same ancestry categories (see Table 1.1), the researchers used the tables and maps published by each of the three companies to recategorize each one's reported ethnicity into 13 common ethnicity categories across companies. The concordance of ancestry results when twins were compared by the same company were found to be relatively high, with mean percentage agreement ranging from 94.5% to 99.2%. In contrast, the concordance of ancestry when each participant was tested by different companies was lower, with mean percentage agreement ranging from 52.7% to 84.1%. The researchers suggested, as I have done here, that the differences across different companies could be explained by the different reference groups used by each company. They concluded their article by noting: "By providing detailed numerical results accompanied by color-coded maps showing where descendants came from, testing companies create the impression of rigor and precision. Our results raise questions about the consistency of both ancestry and trait genetic testing. Consumers should not be misled about the usefulness of these services."[18]

These results clearly show that one should be careful in the interpretation of DNA ancestry testing results. As genealogist Blaine Bettinger wrote: "Unfortunately, ethnicity prediction is still a young and developing science, and ethnicity estimates are subject to limitations that minimize—but do not completely negate—their applicability to genealogical research." As he correctly pointed out, what is informative are not the particular percentages, but the overall patterns.[19] There are two issues related to these limitations that I consider throughout the present book. The first one is that the various ethnicity categories are defined based on specific assumptions about sampling and presumed ancestry. The second issue is that the estimates that these companies provide customers with are the outcome of statistical calculations that are based on models that in turn depend on particular assumptions. This is very important because the correct interpretation of the DNA ancestry testing results has to take into account these two issues.

Let us now see how people understand and interpret the results of DNA ancestry testing.

---

[18] Huml, A. M., Sullivan, C., Figueroa, M., Scott, K., and Sehgal, A. R. (2020). Consistency of direct-to-consumer genetic testing results among identical twins. *The American Journal of Medicine*, *133*(1), 143–146.
[19] Bettinger, B. T. (2019). *The family tree guide to DNA testing and genetic genealogy* (2nd ed.). Cincinnati, OH: Family Tree Books, pp. 173, 177.

Took the Test—Now What?

In 2015, sociologist Rogers Brubaker pointed out that attention regarding ancestry had already shifted from continental-level categories corresponding to human races, to more specific, smaller-scale categories corresponding to ethnic groups, nations, and particular regions. He predicted that this shift was likely to continue with two possible outcomes. On the one hand, it might contribute to naturalizing ethnic and national categories, and even others at lower levels, in the case of people who would be assigned to one such category. On the other hand, it might contribute to undermining notions of pure and discrete racial, national, or ethnic categories, in the case of people who would be simultaneously assigned to multiple categories. Brubaker concluded: "No doubt both processes will occur; genetic genealogy takes many forms, and its effects are likely to be contradictory and ambiguous."[20] Well, he could not have been more correct. This is why it is interesting to consider the conclusions of some studies about how people interpret their DNA ancestry testing results.

Initially, studies focused on people's perceptions of and attitudes toward the tests. Sociologist Jo Phelan and colleagues aimed at testing two hypotheses, with respect to admixture tests, that is, tests that can assign more than one ancestry to a participant. The "reification hypothesis" suggested that the methods of the tests reify race as a genetic reality by implying that racial groups are fundamentally different so that these differences can be identified by the tests. In contrast, the "challenge hypothesis" proposed that receiving admixture test results may challenge, rather than reify, racial categories, as these results generally report that people have mixed racial backgrounds. The researchers provided participants with a vignette containing statements that represented both of these hypotheses, and compared their effect to that of two other vignettes: the "race as social construction" vignette that emphasized broad genetic similarities between racial groups, and the "race as genetic reality" vignette that emphasized broad genetic differences between racial groups. The results clearly supported the "reification" hypothesis. The mean belief in essential racial differences was not significantly different for the participants who read the "admixture" vignette and those who read the "race as genetic reality" vignette. In contrast, the mean belief in essential racial differences was significantly lower among participants who read

---

[20] Brubaker, R. (2015). *Grounds for difference.* Cambridge, MA: Harvard University Press, p. 74.

the "race as social construction" vignette and those who read no vignette at all. The researchers concluded by noting: "our findings point to the possibility that an unintended consequence of the modern genomic revolution is to magnify the degree, generality, profundity, and essentialness of the racial differences people perceive to exist."[21]

Another study was conducted by political scientists Jennifer Hochschild and Maya Sen. They focused on African Americans because they considered them as more likely to be interested in DNA ancestry testing than others, as they might not know much about their past because of the transatlantic slave trade (see Chapter 3). To test the hypothesis that African Americans are more interested in, and influenced by, DNA ancestry tests than others, the researchers developed four vignettes featuring fictitious characters who had just received the results of a DNA ancestry test. Two vignettes (one for many races and one for a single race) asked participants to imagine that they were the vignette individual, whereas the other two (again, one for many races and one for a single race) asked participants to imagine how the individual depicted in the vignette would feel after receiving test results. Overall, the results supported the researchers' initial hypothesis: 45% of African American people agreed with the statement that the test results would "matter a lot" to identity, compared to 41.5% of Hispanic people, 36.5% of Asian American people, 32.5% of multiracial people, and 24.5% of White people. As they noted "People want to forge some of the links in the broken chain of their heritage, even if the result is not exactly what they had hoped for."[22] Such studies with hypothetical vignettes have limitations, of course, but it is interesting to consider their findings. However, the really interesting results are those in the studies with actual test-takers.

Sociologist Wendy Roth has conducted several studies with people who have taken DNA ancestry tests. One of those had two parts: an online survey with individuals anywhere in the world who had taken a DNA ancestry test and subsequent telephone interviews with some of them. Test-takers answered questions about their racial and ethnic identities before taking the test, how these changed after receiving the results of the test, and how these results affected their sense of identity, their activities, and their

---

[21] Phelan, J. C., Link, B. G., Zelner, S., and Yang, L. H. (2014). Direct-to-consumer racial admixture tests and beliefs about essential racial differences. *Social Psychology Quarterly, 77*(3), 296–318.

[22] Hochschild, J. L., and Sen, M. (2015). To test or not? Singular or multiple heritage? Genomic ancestry testing and Americans' racial identity. *Du Bois Review: Social Science Research on Race, 12*(2), 321–347.

18 ANCESTRY REIMAGINED

friendships. I should note that Roth and colleagues considered the division into races as being based on biological characteristics, whereas the division into ethnicities as being based on shared ancestry, history, and culture. The results? Overall, 61% of participants reported that their ethnic or racial identities did not change after testing; in contrast, 19.3% reported that they had thought of their race differently after the test, 34.7% said the same for ethnicity, and 14.9% said this for both. The only group that differed significantly from this average were those who had initially identified as Native American only, with three-quarters of them experiencing an identity change both for race and for ethnicity.[23]

Subsequently, some of these people participated in two rounds of telephone interviews. The first interview focused on their ethnic and racial identity. Overall, 36% of participants stated that they underwent an identity change due to their DNA test results. Interestingly, 59% of participants maintained their previous identities despite receiving new ancestry information. In contrast, 7% of participants considered their results to confirm their preexisting identity. Among those who had identified as White only before testing, 51.6% experienced an overall identity change, whereas among those who had identified as Black only before testing, 16.7% experienced an overall identity change; all the other groups fell somewhere in between. In general, participants experienced an identity change with respect to race slightly more than ethnicity (22% and 16%, respectively). Finally, it is interesting to note that among those whose identities changed, most did not give up their previous identities but rather incorporated additional ones. The most interesting conclusion from the interviews was that test-takers did not simply adopt what the test results stated. Rather, they made choices with the aim to embrace only those identities "that offer distinctiveness and provide social or psychological value," while weighting the social cost of either rejecting or questioning these choices.[24]

The identity changes described in this study were related to change from a particular ancestry category to one or more others (I consider the notion and perception of identity in the next chapter). Details notwithstanding, the implications of such an identity change can be enormous. In the Preface we

[23] Roth, W. D., and Lyon, K. (2018). Genetic ancestry tests and race: Who takes them, why, and how do they affect racial identities. In K. Suzuki and D. A. von Vacano (Eds.), *Reconsidering race: Social science perspectives on racial categories in the age of genomics*. New York: Oxford University Press, pp. 133–170.

[24] Roth, W. D., and Ivemark, B. (2018). Genetic options: The impact of genetic ancestry testing on consumers' racial and ethnic identities. *American Journal of Sociology, 124*(1), 150–184.

saw that Karen and Jay were surprised, if not shocked, by the DNA ancestry test results they received. Before taking the tests, they thought that they had a single ancestry: Karen considered herself East African, and Jay considered himself English. But the results indicated otherwise. Instead of being East African, Karen was told she was mostly Southeast African; instead of being English, Jay was told he was mostly Irish. Furthermore, they were both told that they also had ancestries from other regions. So, instead of confirming the "pure" ancestry that Karen and Jay thought they had, the DNA ancestry test results revealed a different ancestry story. If it were a picture, their ancestry would look more like a colorful patchwork, rather than a uniform whole. These results mattered to Karen and Jay because they affected their perception of who they are. If they were not who their parents had told them they were, then who were they? So in these two cases the results were rather against an essentialist view of social categories such as race. Simply put, an essentialist view of race perceives the latter as due to inner essences that result in distinct and internally homogeneous categories (see Chapter 2 for a detailed discussion of essentialism and Chapter 3 for essentialism and race).

Roth and colleagues designed another study in order to investigate if the results of these tests increase essentialist views of race. Half of the participants received Admixture and mtDNA tests purchased from Family Tree DNA, and half received no test. These participants were asked what they thought about the relationship between DNA and race, both before and after taking the test.[25] Overall, the results showed no significant average effect of testing on essentialist views about race between the group who took the test and those who did not. However, when the researchers looked at the effect with respect to participants' knowledge of genetics, there were differences. Among those who took a test, participants with high genetics knowledge showed a significant decline in their essentialist views of race after taking the test, whereas they also had lower genetic essentialism beliefs before taking it. In contrast, these beliefs increased among those with very limited knowledge of genetics. However, due to their limited numbers, the researchers suggested that this requires further research. There was also no change in essentialist beliefs with respect to participants' specific ancestries. Therefore, a main conclusion was that participants' genetics knowledge may be critical

---

[25] Yaylacı, Ş., Roth, W. D., and Jaffe, K. (2021). Measuring racial essentialism in the genomic era: The genetic essentialism scale for race (GESR). *Current Psychology*, *40*, 3794–3808.

20 ANCESTRY REIMAGINED

for how they understand the results of DNA ancestry tests—but we should keep in mind that the study only included people who identified as White.[26]

Despite this limitation, I think that the findings of this study are worth considering as they clearly point to an important direction for future research. It would thus be very interesting to investigate, with DNA ancestry test-takers from as many social groups as possible, whether increasing the level of genetics understanding would result in decreasing racial essentialism. I am inclined to think that this would be the case for non-White people as well, given the findings from a similar science education research program led by science educator Brian Donovan. Those studies included students from majority-White schools, which also enrolled students who identified with several others US Census categories besides White. It should be noted that Donovan and colleagues tried to recruit as a diverse sample as possible. As they wrote in their article, "We attempted to recruit twice as many schools, especially those serving lower income students, more racially diverse populations, or those located in more politically conservative areas. However, district research offices for these schools rejected our study." Despite this, I think that their results are important, and certainly point to useful future research directions.[27] A better understanding of genomics, which included how genetic essentialism views are not supported by the patterns of human genetic variation, also resulted in greater reduction in the perception of racial differences and in belief in genetic essentialism.[28] Whether genetics and genomics education can contribute to reducing racial biases and essentialism is a research question certainly worth investigating further.

[26] Roth, W. D., Yaylacı, Ş., Jaffe, K., and Richardson, L. (2020). Do genetic ancestry tests increase racial essentialism? Findings from a randomized controlled trial. *PloS One, 15*(1), e0227399.

[27] To test their hypotheses, the researchers used three different randomized control trials (RCTs), two of which with school students. The sample of 166 middle and high school students in RCT1 self-identified as White (48%), Mixed-Race (19.88%), Asian (18.1%), Hispanic (5.42%), and Black and Pacific-Islander (1.2% each). The remaining students did not select one of the US Census categories (6.02%). The 721 high school students in RCT3 self-identified their race as White (61.7%), Asian (19.8%), Mixed-Race (9.9%), Hispanic (4.9%), Black (2.4%), Pacific-Islander (0.55%), and American Indian (0.4%). See Donovan, B. M., Semmens, R., Keck, P., Brimhall, E., Busch, K. C., et al. (2019). Toward a more humane genetics education: Learning about the social and quantitative complexities of human genetic variation research could reduce racial bias in adolescent and adult populations. *Science Education, 103*(3), 529–560.

[28] Donovan, B. M., Weindling, M., Salazar, B., Duncan, A., Stuhlsatz, M., and Keck, P. (2021). Genomics literacy matters: Supporting the development of genomics literacy through genetics education could reduce the prevalence of genetic essentialism. *Journal of Research in Science Teaching, 58*(4), 520–550.

But this would not necessarily be the case for everyone. Particular groups of people can selectively draw on and interpret scientific research to justify their politics. Sociologist Aaron Panofsky and colleagues investigated how White nationalists interpret the results of DNA ancestry tests. There are several discussions about how White supremacists could "prove" their racial purity with DNA ancestry testing; and about how their opponents could rely on these tests to undermine their claims. Recall Craig Cobb who received, and quickly dismissed, the test results that he had 14% African ancestry. What the researchers did was to analyze posts from the White nationalist website Stormfront.org, in which users revealed the results of their own DNA ancestry tests. The aim was to document the reactions of the community to results that did not confirm White racial purity. The most important finding was that the reactions of the community were mostly about the interpretation of the tests' results, rather than about the users posting them. The Stormfront community was generally devoted to address a user's upsetting results, rather than criticizing them. This was done through rhetorical strategies that either rejected the results or reinterpreted them. Another interesting finding was that the community's reactions depended more on the attitude of the user who made the announcement than on the results themselves. Those users who appeared to be humble were more likely to receive helpful messages than those who were extremely defensive.[29] The important issue here is that, contrary to arguments that learning about mixed ancestries may reduce racial prejudice, there is evidence that racial supremacists can adjust their dogmas in a way that makes them compatible with DNA evidence. In this way, the results of DNA ancestry testing can be selectively invoked for political purposes.

Another example of the political use of DNA ancestry testing results comes from Canada. Sociologist Darryl Leroux has explained how a minority of White French descendants living in Canada have tried to establish their Indigenous identity by showing through genealogical research that they have had an Indigenous ancestor between 300 and 375 years ago. Whereas it is demographically established that only a few Indigenous women were married to French colonists, claims to Indigenous identities have been made by the descendants of these French colonists based on a very small (less than 1% on average) Indigenous ancestry, documented via DNA ancestry

---

[29] Panofsky, A., and Donovan, J. (2019). Genetic ancestry testing among white nationalists: From identity repair to citizen science. *Social Studies of Science*, 49(5), 653–681.

22 ANCESTRY REIMAGINED

testing. As Leroux pointed out, this has led to a change in identity claims among White people and to the expansion of whiteness in ways that were not conceivable just a few decades ago. In some cases, claiming an Indigenous identity is a means to oppose Indigenous land claims. Whereas it is not really possible for a DNA ancestry test to establish one's Indigenous origins, these people took a test by a company called Accu-metrics that claimed to be able to indicate a person's origins among over 600 tribes or First Nations in the United States and Canada, as well as 56 Native tribes from Mexico. Based on these tests, the descendants of French colonists dramatically redefined their origins, claiming a 2,000-year-old origin in the area, even though their history places them there no more than 300 years ago. In this case, there was a conscious shift in identity for political purposes. Interestingly, Leroux himself took the Accu-metrics (mtDNA) tests. The results he received were a 9% Native American ancestry, which would correspond to between 150 and 200 Indigenous ancestors in the early 1600s; but having been able to identify through genealogical methods all of his ancestors up to that time, only three of whom could be identified as Indigenous, Leroux concluded that the test grossly miscalculated his ancestral story.[30]

Given the findings presented in this section, I think that there are two important ways forward. The first is to confirm that genetics and genomics education can help reduce racial essentialism among test-takers, and among the public in general. If this were found to be the case, robust teaching interventions might help reduce biases. The second is to create the opportunities for scientists working in genetics and genomics to get more involved in science communication and explain why our current knowledge of human DNA variation does not support any notion of biological race (see Chapters 7 and 8).

Where Does All This Take Us?

This chapter has presented some research findings that I summarize here:

- Most DNA test-takers initially self-report a single ancestry identity. This identity is usually the predominant one, but not the only one, in their DNA test results.

---

[30] Leroux, D. (2019). *Distorted descent: White claims to indigenous identity.* Winnipeg: University of Manitoba Press, pp. 1–4, 202–210.

- Different companies can provide different results for the same person, which depend on the reference groups and the algorithms used in each case.
- Monozygotic twins should be expected to receive identical DNA ancestry reports, and they usually do.
- People generally think that their ancestry can be determined by a DNA test, either by confirming their initial identity or by leading them to reconsider it.
- Most DNA test-takers do not passively accept the results, but actively embrace them to fit in their broader life stories.
- Overall, DNA ancestry results can affect people's self-perception in different ways. Results about mixed ancestry might affect one's self perception, but it is also the case that people find ways to either reject or accommodate their results.

So far, so good. But then, you might wonder, what is the rest of this book? Well, by the end of it you will have realized that understanding ancestry and DNA ancestry testing is a bit more complicated than that. But to get there, we first need to understand the concept of ancestry (Chapters 2–5) and realize some paradoxes underlying it (Chapters 6–9). These paradoxes, each considered in a whole chapter, are as follows:

I. Ancestry is collective and global, but we think of it primarily as individual and local.
II. Ancestry is mostly about relatedness, but we privilege distinctiveness.
III. Ancestry groupings are human inventions, but we consider them as natural.
IV. Ancestry depends on culture, but we think of it as based on DNA only.

Only after understanding what ancestry is and is not, and only after considering these four paradoxes about it, I argue, can we really understand what DNA ancestry testing can, and cannot, "tell" us.

2

Essentializing Social Groups

Nations

"Who Am I?": Identity and Belonging

There are several different answers one can provide to the question "Who am I?" For instance, I could say that I am a man; a parent; a son; a Greek. But I am not the only person who has these attributes, as there are many others who are men, parents, sons, and Greeks. Specific features can further be used to differentiate us from one another: Whose parent? Whose son? From which part of Greece? And there are also other people who do not have any of these attributes: just think of a woman who has no children, who is someone's daughter, and who is Swiss. These attributes considered together can prescribe a person's identity, which makes each one of us unique and distinct from the other people around us. At the same time, each one of these attributes can be considered as an identifying feature, and therefore a specific identity, of a person: one can be a biologist or a physicist; a fan of the Boston Celtics or the Los Angeles Lakers; a Democrat or a Republican; a Muslim or a Christian; and so on.

Identities matter both for how we perceive ourselves, as well for how others perceive us. These perceptions can, in turn, affect out behavior toward one another. As philosopher Kwame Anthony Appiah has noted, once a person is assigned to a particular identity, this determines both how a person treats other people and how a person is treated by other people. When it comes to ancestry, the identities that matter are those related to race, nationality, or ethnicity. On the one hand, there is the feeling of solidarity with members of the same group because their common (racial, national, or ethnic) identity is perceived to entail that they should care about and support one another. For instance, meeting a compatriot abroad often makes people feel closer to them than to people of other nationalities. On the other hand, there is the feeling of estrangement, and sometimes of hostility, for people who have identities that are considered to be antithetical to one's own. Just think of two

Ancestry Reimagined. Kostas Kampourakis, Oxford University Press. © Oxford University Press 2023.
DOI: 10.1093/oso/9780197656341.003.0002

people whose nations are in conflict (e.g., Russia and Ukraine). In short, you may help or be helped, as well as hate or be hated, based on your (racial, national, or ethnic) identity.[1]

The people who share a particular identity can be thought of as forming a group to which they belong. However, belonging to a group is not just about membership but also about an emotional attachment to that group. This emotional attachment can transform individual identities so that they acquire features of the collective identity of the group. An individual who belongs to a group is expected to be loyal to it and follow its values and principles—the price to pay for enjoying the rights of being a member of the group. Belonging also entails a sense of familiarity, as the group offers a "home" wherein individuals share common interests and values. The feeling of belonging is further enhanced by the attachment of the group to a homeland. The individual is therein identified as "one of us" and thus considered important. In short, there is a reciprocal commitment between the individual and the group. As political scientist Montserrat Guibernau has shown, it is for all these reasons that belonging matters: it offers individuals emotional warmth and a feeling of security.[2] Finally, belonging is always accompanied by exclusion, as it immediately raises the distinction between members and strangers, between "us" and "them." Sociologist Nira Yuval-Davis has suggested that the boundaries between distinct groups are defined and maintained by the politics of belonging, which determine the processes that construct these boundaries, decide about the inclusion or exclusion of particular people, and define what belonging entails for the members of the group. These processes are guided by the political agents that have the power to make such decisions. The criteria about inclusion and exclusion can be common descent, place of birth, culture, religion, or language.[3]

Among the three different kinds of identities related to ancestry—race, nationality, ethnicity—it is nationality that is easier to assign because it has an official status, with official documents such as an identity card and a passport, and with formal rights, such as those related to residence and work, and obligations, such as those related to military service and taxes. So let us consider by way of example what belonging to a nation means. The members

---

[1] Appiah, K. A. (2018). *The lies that bind: Rethinking identity*. London: Profile Books, pp. 8–12.
[2] Guibernau, M. (2013). *Belonging: Solidarity and division in modern societies*. Cambridge: Polity, pp. 28, 30, 32–34.
[3] Yuval-Davis, N. (2011). *The politics of belonging: Intersectional contestations*. Los Angeles: Sage, pp. 10, 15, 18–21.

26 ANCESTRY REIMAGINED

of a nation can share a homeland, a history, a distinct public culture, and customs. These shared features provide a sense of unity among themselves and a sense of difference from those perceived as outsiders. Belonging to a nation also creates an emotional attachment to it and a commitment to its values. These may sometimes contradict the individual's personal values, for instance, by obliging a pacifist to do a military service. The emotional attachment becomes most evident in those cases where people are willing to sacrifice themselves, in order for their group and everything related to it to continue to exist—as in the case of a war (recall from the Preface how proud Jay was about his ancestors fighting for England). This emotional attachment is enhanced by a drive for homogenization during the processes of nation-building, such as the commemoration of particular historical events, the way national history is taught at schools, or the criteria for immigration and naturalization. But perhaps the most important feature is the specification of national borders for the homeland, which tend to unify the members of a nation both in peace and in times of crisis. The idea of protecting a nation's borders is further enhanced by the notion of indigeneity, which entails that the members of a nation belong to the homeland and have the most authentic rights to make such a claim.[4]

I should note here that indigeneity is a tricky concept. It usually refers to someone originating in a particular place. But at what moment in time are origins found? Think for instance of the United States. Today, Indigenous people are considered those who are the descendants of those peoples who lived in the American continent before the European colonization. As their descendants are a minority of the US population today, the United States is often described as a country of immigrants. Many of the descendants of these immigrants also make claims of indigeneity when they compare themselves to more recent immigrants (e.g., think of former US President Donald Trump, himself a second-generation immigrant from Europe, who wanted to build a wall on the borders with Mexico in order to prevent immigrants from entering the United States). Many people whose recent ancestors were immigrants died in wars fought by the United States, believing that they were defending their homeland in which they were born and raised. So, if these people were born and raised in this nation, and all felt a sense of belonging to it, shouldn't they also be considered Indigenous? I do not want to propose

---

[4] Knott, E. (2017). Nationalism and belonging: Introduction. *Nations and Nationalism, 23*(2), 220–226; Yuval-Davis, N. (2011). *The politics of belonging: Intersectional contestations*. Los Angeles: Sage, pp. 81–112.

ESSENTIALIZING SOCIAL GROUPS 27

any answer to this question here, but only to highlight that some concepts are trickier than what we usually think. Who is considered Indigenous in a nation is mostly a matter of social convention. Indigeneity is especially important when it comes to ancestry, because it is Indigenous people who we consider as having ancestry from a particular region. Therefore, how we define indigeneity matters.

Indigenous studies scholar Kim Tallbear has argued that a Euro-centric perspective has guided research on Native American (and I would say Indigenous DNA all over the world more broadly), on the basis of two justifications. The first one has been that studying the DNA of Indigenous people was useful for the sake of knowledge. But as a result of this, the concerns and the perspectives of the Indigenous people involved have often been overlooked, and they have been often left out of any discussion about the aims and the methods of such studies. The second justification has been that because Indigenous peoples might vanish due to genetic admixture (a highly problematic concept that I consider in Chapter 8), it is absolutely urgent to study their DNA before it is too late.[5] Such have been the justifications for various human genome diversity projects (some are mentioned in Table 6.1), which have often taken place without consulting the Indigenous peoples sampled. But this has raised crucial ethical issues. For instance, DNA has come to be a point of reference for race, identity, and belonging, with implications for Indigenous North Americans' sovereign rights to their lands and heritage. As a result, Indigenous peoples have raised concerns about the lack of consultation and community engagement in many studies of Indigenous ancestors, about how permissions are obtained for research, about the control and dissemination of data, as well as for its potential misuse and commodification. To some, the view that human remains are a historical resource to be mined is seen as another form of colonialist violence against Indigenous peoples. As a result, a number of Indigenous communities have banned DNA research or demand greater control of both the process and products.[6]

Returning to identity and belonging, the people who share the same national identity are usually perceived to share some common features, such

---

[5] TallBear, K. (2013). *Native American DNA: Tribal belonging and the false promise of genetic science*. Minneapolis: University of Minnesota Press, p. 2.

[6] Cortez, A. D., Bolnick, D. A., Nicholas, G., Bardill, J., and Colwell, C. (2021). An ethical crisis in ancient DNA research: Insights from the Chaco Canyon controversy as a case study. *Journal of Social Archaeology, 21*(2), 157–178.

28 ANCESTRY REIMAGINED

as language and culture. Because of this, national identity provides a basis for stereotyping. On the one hand, it is assumed that people of the same nationality are a lot more similar to one another than they actually are. On the other hand, it is assumed that people of different nationalities are much more different from one another than they actually are. Therefore, once considered as a member of a particular nation, a person is no longer considered as an individual, but as a representative of particular national type. This results in people not being judged on the basis of who they really are, but rather on the basis of who they are expected to be based on these stereotypes. To give some examples, being Greek has been for some northern Europeans the synonym for being lazy and unreliable; for others, it is the synonym of being openhearted and hospitable. Similarly, being German has been considered by some southern Europeans as the synonym for being insensitive and cold; for others, it is the synonym of being disciplined and reliable. Such stereotypes can easily prevail; perhaps because they make life easier as one does not need to think.[7]

But if we do think about it, national identity is only one of the many different identities we simultaneously have, which are not independent but rather influence one another. Therefore, people who share one identity may not have similar experiences due to the influences of their other identities that they do not share—an idea called intersectionality by Kimberle Crenshaw, a law, race, and feminism scholar. For instance, one might consider the intersections of gender and race. Black women, for example, often find that they face discrimination against their Blackness and against their gender; thus, their experiences are both similar and different from those of White women, or of Black men.[8] Intersectionality is therefore the reason why it is difficult to arrive at generalizations about nationalities (or races, or ethnicities for that matter). People who share one identity may have different experiences and different social issues to deal with due to their other identities. Compare a Jewish person who is White and a US citizen living in Boston to a Jewish person who is White and an Israeli citizen living in Jerusalem. Compare any of these two to a Lemba Jewish who is Black and a Zimbabwean citizen. Or compare the latter to another Black Zimbabwean citizen who is a Protestant Christian. And then compare that person to a

---

[7] Coulmas, F. (2019). *Identity: A very short introduction*. Oxford: Oxford University Press, p. 36.

[8] Crenshaw, K. (1989). Demarginalizing the intersection of race and sex: A black feminist critique of antidiscrimination doctrine, feminist theory and antiracist politics. *University of Chicago Legal Forum*, *1*(8), 139–167.

ESSENTIALIZING SOCIAL GROUPS   29

White Christian Protestant who is Irish. And then the latter to a White Irish Christian Catholic. And so on. My point here is that, once we start thinking deeply about these issues, it is not simple to talk about Zimbabweans or Irish or Jewish or Protestants or Catholic as clear-cut identities, because people often have many different identities at the same time. This is an important point we can easily miss.

What is also important is that even though our different identities are ontologically distinct, we cannot make meaning of them independently because they are "mutually constitutive"; that is, each exists with relation to the others, and they continuously affect each other interdependently. This always takes place in a specific historical moment and context. For instance, to draw on the examples above, the Jewish cultural identity will be experienced very different depending on whether a person lives in Boston, Jerusalem, or Zimbabwe. However, and at the same time, it is possible that in specific contexts and situations some of these identities are more influential than others in shaping people's lives and their positioning with respect to people around them.[9] Therefore, for instance, in spite of having multiple identities, we might prioritize our national identity and belonging. Academic and politician Michael Ignatieff has argued that nationalists consider national belonging as more important than other forms of belonging (e.g., to family, work or friends), because none of these other forms is secure unless there is a nation to protect a person from violence. According to Ignatieff, this is what makes nationalism persuasive. This also makes the nation irreplaceable; even for today's globalized societies and cosmopolitan citizens: "a cosmopolitan, post-nationalist spirit will always depend, in the end, on the capacity of nation states to provide security and civility for their citizens."[10] A cosmopolitan life is possible, according to this view, only if nations have the capacity to provide the security and the rights necessary for everyday life.

It is important to note at this point that, generally speaking, different conceptions of a nation exist. Despite their variety, we can usefully distinguish between two conceptions of nation:

- *Ethnic nation*: In this view, a nation consists of people who share a common ancestry and who have thus inherited, not chosen, their

---

[9] Yuval-Davis, N. (2011). *The politics of belonging: Intersectional contestations.* Los Angeles: Sage, pp. 7–9.

[10] Ignatieff, M. (1994). *Blood and belonging: Journeys into the new nationalism.* London: Vintage, pp. 6 and 9.

30 ANCESTRY REIMAGINED

membership in the nation. Ethnicity and language are very important in this case, as they are among the main criteria for membership. Israel and Greece could be examples of ethnic nations, if they really are.

- *Civic nation*: In this view, a nation consists of people who subscribe to its political creed and obey its laws, independently of their characteristics, such as ethnicity and language. It is called civic because it is envisaged as a community of citizens with equal rights that share a set of practices and values. The United States and Canada are typical examples of civic nations.[11] However, if one looks at human migrations throughout history, it is most likely to conclude that all nations are in fact civic.

For some people, however, it is the ethnic conception of a nation that might seem more intuitive and more natural. Essentialist thinking provides the grounds for this, and it is to this topic that we now turn.

DNA as Essence

When a friend tells you that she bought a new cellphone or a new car, you understand what that is and are very unlikely to confuse one with the other. The reason is that cellphones, cars, and any other object have a set of properties that are characteristic of them and that together specify what each object is. Cellphones are small, have a screen and a battery, and are used to communicate. Cars are large, have four wheels and an engine, and are used to commute. Of course, cellphones are not the only objects that have batteries and screens, and cars are not the only objects that have wheels and engines. But thinking about all their specific features together allows us to distinguish among cellphones, cars, and any other objects. This way of thinking also applies to organisms. When a friend tells you that she has a dog or a cat, you understand what she is talking about. Even if it is not easy to articulate the set of features that are specific to cats and dogs—after all, they are mammals and so they all have legs, hearts, lungs, and so on—they also have distinct features that are specific to them: for instance, the muzzle of dogs or the ability of cats to hold their tails vertical while walking. Therefore, to most people familiar

[11] Ignatieff, M. (1994). *Blood and belonging: Journeys into the new nationalism.* London: Vintage, pp. 3–4; Smith, A. D. (2010). *Nationalism.* Cambridge: Polity, pp. 43–46; Fenton, S. (2010). *Ethnicity* (2nd ed.). Cambridge: Polity, p. 52.

with these animals, it should be clear what a dog or a cat is and how they differ from other animals.

This way of intuitive thinking, according to which a category has specific features shared by all its members, which in turn make them what they are, is described as essentialism. It develops early in life and seems to be actively constructed rather than passively received. Essentialist thinking can be very useful as it helps us assign unknown entities to categories and, based on that, to make inferences about their properties. For instance, dolphins may look like fish, but they are mammals; once we realize this, we can make several inferences about their features that we cannot see, such as that they have lungs instead of gills.[12]

Essentialism consists of a set of interrelated beliefs:

I. Categories consist of fundamentally different kinds of entities (objects, animals, or whatever).
II. The boundaries between categories are strict and absolute, so something either belongs to a particular category or not.
III. Categories are homogeneous; that is, their members share fundamental similarities with one another. At the same time, they have fundamental differences from members of other categories.
IV. These similarities and differences are due to causes within the members of the category that make them what they are.[13]

If we apply essentialist thinking to human social categories, such as nations, we get the following set of interrelated beliefs:

I. Nations consist of fundamentally different kinds of people.
II. The boundaries that separate nations are strict and absolute, in the sense that a person either belongs to a particular nation or not.
III. Nations are homogeneous; that is, their members share fundamental similarities with one another. At the same time, they have fundamental differences from members of other nations.
IV. These similarities and differences are due to inherent causes and make the members of each nation what they are.

[12] Gelman, S. A. (2003). *The essential child: Origins of essentialism in everyday thought.* Oxford: Oxford University Press.

[13] Rhodes, M., and Kelsey, M. (2020). What is social essentialism and how does it develop? In M. Rhodes (Ed.), *The development of social essentialism* (pp. 1–30). San Diego: Academic Press.

32 ANCESTRY REIMAGINED

These inherent causes can be anything inside an individual, but DNA has often been considered as an appropriate placeholder for these. I call this idea *DNA essentialism*.[14] It should be obvious that DNA essentialism is compatible with the ethnic view of nations and incompatible with the civic view. Therefore, depending on how people conceptualize their own nations, they might be more or less prone to essentialize them (see the last section of this chapter). DNA essentialism would likely make sense for people who hold an ethnic conception of nation, based on common ancestry and inherited membership. I consider the assumptions of DNA essentialism about human social groups (nations, races, ethnic groups) in detail in Chapters 3 and 6–9. For now, it suffices to note that the key question when it comes to DNA is whether its variation (see Box 2.1) among members of the same group allows for their clear distinction. This question is addressed in Chapters 6–9.

Let's now think again about the stereotypes considered in the previous section, while considering points I–IV above and applying them to stereotypes such as "Greeks are hospitable/lazy" or "Germans are disciplined/cold." These stereotypes can be taken to mean that a Greek cannot be disciplined/ cold or that a German cannot be hospitable/lazy (I); that one can be Greek or German, but not both (II); that all Greeks are hospitable/lazy and that all Germans disciplined/cold (III); and that there is something within people ("national" DNA?) that makes them who they are. But, of course, none of these is true. Yet, once we hear that someone is Greek or German (or whatever else), it is possible to make inferences about them based on their nationality, without knowing them at all. In other words, through stereotypes we essentialize nations.

Essentializing social groups such as nations seems to serve two distinct roles. On the one hand, it serves to define who "we" are. If the members of a group can specify a set of characteristics that they consider distinctive of themselves, this can enhance their sense of being similar to one another and

---

[14] A related idea is genetic essentialism: the idea that genes are our deep essences. In this view, genes are fixed entities, which are transferred unchanged across generations and which are the essence of what we are by specifying characters from which their existence can be inferred. When genetic essentialism is applied to social categories, such as gender, race, nationality, or ethnicity, one can infer that the members of any of these groups are more similar to one another, and more different from the members of other groups, at the level of DNA. See Heine, S. J. (2017). *DNA is not destiny: The remarkable, completely misunderstood relationship between you and your genes*. New York: WW Norton & Company; Heine, S. J., Dar-Nimrod, I., Cheung, B. Y., and Proulx, T. (2017). Essentially biased: Why people are fatalistic about genes. In J. M. Olson (Ed.), *Advances in experimental social psychology* (Vol. 55, pp. 137–192). San Diego: Elsevier Academic Press. DNA essentialism differs from genetic essentialism is that it is not restricted to genes. Ancestry inferences are based on DNA sequences that can but do not have to be (and usually are not) within genes.

Box 2.1 What Is DNA Variation?

The sequence of DNA molecules (see Box 1.1) can change, either due to mistakes during the process of their production, called DNA replication, or due to environmental factors that can cause changes to them—one of which is ultraviolet radiation. Even though cells have proteins that can "correct" many of these changes, some of them are not corrected, thus resulting in changes in the DNA sequences that are described as mutations.[1] This term simply means "change." Some of these changes can bring about a change in a characteristic; this usually happens when mutations occur within genes or sequences related to their function. When changes in DNA occur elsewhere, they usually have no effect on the organism and they simply exist there. These are usually described as single nucleotide polymorphisms (SNPs; *polymorphism* literally means "many forms"). SNPs usually emerge when a certain nucleotide is replaced by another (e.g., A→G). The study of human DNA variation can be based on the comparison of which SNPs different individuals have. The SNPs that are used for comparison in ancestry studies are usually dimorphic (which means that they exist in two forms each). Besides SNPs, there exist some other variable DNA elements in our genomes. These are short or longer repeated sequences, and what differs among individuals is the number of repeats that one has. These can be generally described as copy-number variants (CNVs) and their mutation rate, that is, the frequency of mutations occurring to them over time, is much higher than that of SNPs.[2]

[1] It has been estimated that a newborn child of 30-year-old parents will on average carry 75 new mutations. Besenbacher, S., Sulem, P., Helgason, A., Helgason, H., Kristjansson, H., Jonasdottir, A., . . . Stefansson, K. (2016). Multi-nucleotide de novo mutations in humans. *PLoS Genetics*, *12*(11), e1006315.

[2] Jobling, M., Hollox, E., Hurles, M., Kivisild, T., and Tyler-Smith, C. (2013). *Human evolutionary genetics* (2nd ed.). New York: Garland Science, pp. 48, 65–70, 75.

the importance of their membership to that particular group as a source of solidarity and support. On the other hand, essentialism serves to distinguish among different groups and perhaps justify a privilege status of one over another. Once the specific characteristics of a group are defined, it is possible to distinguish, and perhaps discriminate, between the members of our ingroup (the group to which we psychologically identify as being members) and the

34 ANCESTRY REIMAGINED

members of outgroups (other groups to which we did not identify).[15] In short, essentialism makes us prone to distinguish between "us" and "them," and this can enhance both the cohesion of the ingroup and the exclusion of outgroups. This is why we may feel closer and friendlier to those who belong to the same nation with us, and at the same time feel more distanced from those belonging to other nations.

This is important to keep in mind because the marketing strategies of DNA ancestry companies rely on a rhetoric that might be perceived as being essentialist, and thus consisting of beliefs I–IV above. For instance, on Ancestry.com we read: "You could be Irish. More specifically, Munster Irish. AncestryDNA® doesn't just tell you which countries you're from, but also can pinpoint the specific regions within them, giving you insightful geographic detail about your history."[16] This phrase might be interpreted as being based on particular essentialist assumptions: that "Irish" or "Munster Irish" is a distinct group that can be delimited from others—otherwise there is no point in referring to such a group (I); that a person who is "Munster Irish" cannot also belong to another group, say being "Connacht Irish"—otherwise it is meaningless to say that one is "Munster Irish" if one could also be "Connacht Irish" (II); that there is something distinctive among "Munster Irish" people that makes them more similar to one another and more different from other categories, such as "Connacht Irish"—otherwise they could not be identified as such (III); and because being "Munster Irish" is something that we can find based on DNA—otherwise why should we do the DNA test anyway?—which means that DNA is the essence of who we are (IV).

However, this is not necessarily so. A person can acquire a national (or ethnic) identity from their family independently of their genetic connection to them. Consider, for instance, the story of Jim Collins presented in detail by journalist Libby Copeland in her book *The Lost Family*. Jim grew up in an Irish family, even though his biological parents were Jewish. A mistake that occurred in the hospital in which he was born in 1913 resulted in the exchange between him and another boy born on the same day, so that they ended up in each other's biological family. Jim grew up being Irish, and he also raised his children within the Irish cultural traditions. It was several years after his death that his daughter Alice took a DNA test that attributed

---

[15] Diesendruck, G. (2020). Why do children essentialize social groups? In M. Rhodes (Ed.), *The development of social essentialism* (pp. 31–64). San Diego: Academic Press.
[16] https://www.ancestry.com/dna/?oiid=109972&olid=109972&osch=Web+Property (accessed January 19, 2022).

ESSENTIALIZING SOCIAL GROUPS   35

Jewish ancestry to her. The subsequent research she and her sister undertook over the course of several years revealed the mistake that had happened in the hospital where their father had been born about a century ago. Jim grew up being Irish; the other boy, Philip Benson, grew up as Jewish.[17] Both Jim and Philip's sense of belonging was likely toward the groups of the families in which they grew up, not the biological ones. Being Irish or Jewish or whatever can thus be entirely independent of the DNA one has inherited from their parents.

However, it is also the case that we might rely on DNA to clarify our ancestry, if we do not know much about our ancestors, or if we have conflicting information about them. Many people have thus been investing time and money in the search for their roots, imagining pure and unbroken lineages that connect them to a distant past, and emphasizing personal and ethnic origins. Science studies scholar Venla Oikkonen has suggested that we perceive the connections between the present and the past in an affectively charged way, which she has called *evolutionary nostalgia*. Nostalgia is the sentimental longing for a past situation or condition, which literary means "a pain for return." It relies on an imagined causal connection between the present and the past, as well as on the idea that DNA is a record of evolutionary history. However, as she has argued and as I also explain throughout the present book, the information about ancestry and belonging that people seek does not exist independently of us. Ancestry and belonging are constructed, not discovered. They are constructed on the basis of biological materials, scientific methods, and assumptions, as well cultural imaginaries and historical narratives.[18]

Yet essentialism makes us prone to believe that DNA provides objective criteria for belonging. If DNA is the record of history, then the study of our DNA could show us where we come from and where we belong. By having our DNA analyzed, we become members of particular genetic communities. For instance, Ancestry used to describe these genetic communities as "clusters of living individuals that share large amounts of DNA due to specific, recent shared history."[19] More recently, Ancestry has been referring to "communities" only, not "genetic communities," but

---

[17] Copeland, L. (2020). *The lost family: How DNA testing is upending who we are*. New York: Abrams Press.

[18] Oikkonen, V. (2018). *Population genetics and belonging: A cultural analysis of genetic ancestry*. Cham, Switzerland: Palgrave Macmillan, pp. 132–133, 147, 221, 225.

[19] https://www.ancestry.com/cs/dna-help/communities/whitepaper (accessed December 1, 2021).

what connects the people that belong to them is still found at the level of DNA: "The populations from which communities descend were groups of people who (due to barriers to travel or a tendency to marry within the same religion or ethnic group) shared significant amounts of DNA with one another, making them genetically distinct. Once we identify a community of people who are connected through DNA, we look for patterns in family trees linked to AncestryDNA tests to help us identify who the group consists of."[20] Thus, DNA ancestry testing can provide us with a new sense of belonging to genetic communities, which might or might not correspond to our nation. These are communities consisting of people who have essential similarities (thinking of DNA as our essence) and who are thus essentially different from the people of other communities.

Let us now consider what these genetic communities can be.

Genetic Communities

There are many examples of groups that have been perceived as genetic communities, but the Jewish are probably the most exemplary case. Anthropologist Nadia Abu El-Haj has analyzed in detail the search for biological underpinnings of Jewish identity throughout the 20th century. Despite being dispersed in various regions of the world, Jewish communities have been considered to be unusually endogamous—that is, mating with one another only and not with people from other communities—and to also have a common origin from a "Jewish people" who were exiled from ancient Palestine. In the aftermath of World War II, with the establishment of the state of Israel, Israeli-Jewish scientists began mapping Jewish genetic diversity. This kind of research might provide evidence for the genetic unity of Jewish people. But, if one were to think of Jewish people as a genetic community, there would be two important concerns to address. The first one is whether Jewish people from other places in the world, particularly those described as "oriental Jews," could really be "national kin" with European Jewish people. The second is how to establish that the origin of the European Jews is actually in the Middle East, as their religion prescribes. This would cause a paradoxical situation. On the one hand, a Jewish genetic community would exist

---

[20] https://support.ancestry.com/s/article/AncestryDNA-Communities?language=en_US (accessed March 22, 2022).

if oriental Jews were shown to be genetically similar to European Jews but still distinct from Arabs in the neighboring regions. On the other hand, in order to establish that contemporary Jewish communities are descendants of people that left ancient Palestine, they had to be shown to be genetically related to contemporary Arab populations. In short, Jewish people had to be both genetically similar to and genetically distinct from Arabs in order to confirm the traditional story of Jewish diaspora.[21] But, in fact, they have never been genetically distinct in any absolute sense. As geneticist Raphael Falk has argued:

> We may conclude that there is no unique and unifying "biology of Jews." In other words, although the social and cultural relations among Jewish communities (and the very different relations between each of them and its Gentile [non-Jewish] neighbors) have resulted in the formation of a loosely linked cluster of reproductive isolates, any general biological definition of Jews is meaningless. Despite the persistence of intra-Jewish, sociocultural relatedness, coupled with the exclusion of Jews by Gentile society, Jewish communities have never been reproductively isolated from their neighbors.[22]

Falk, himself a Jewish geneticist, argued that the evidence is in favor of constructed, not genetic, communities. Despite extensive intra-marriage among Jewish people, there has always been intermarriage with neighboring peoples.

Such aspirations are found elsewhere. They may be intentional, but they may also be side effects of the choices of geneticists, both local and foreign, who study particular human populations. Historian Elise Burton has shown that during the 20th century, human population geneticists focused on the Middle East because they considered the supposedly endogamous communities living there as living representatives of the past (and many still do), and thus particularly valuable for the study of human diversity. But in order to establish any genetic distinctiveness of such populations, they had to be compared with something else. Thus, geneticists in the Middle East transformed religious, linguistic, and other social identities into ethnicities,

---

[21] Abu El-Haj, N. A. (2012). *The genealogical science: The search for Jewish origins and the politics of epistemology.* Chicago: University of Chicago Press, pp. 16–19, 44, 52.
[22] Falk, R. (2017). *Zionism and the biology of Jews.* Dordrecht: Springer, p. 208.

38 ANCESTRY REIMAGINED

comparing supposedly endogamous communities such as the Armenians, the Jews, and the Bedouins with other communities in which interbreeding with other groups had occurred, such as the Persians, the Turks, and the Arabs. Furthermore, the use of national labels, such as "Iranian" or "Turkish," to identify populations, even though these labels do not refer to biologically meaningful groups, resulted in a practice of *methodological nationalism*, in which the nation-state was considered unquestionably to be a natural unit of analysis. As a result, genetics research refocused national imagination toward biological continuity, thus supporting the idea of genetically distinct nations. The requirement for genetics research to describe human populations according to geography and ancestral history, which are the main constituents of the nation, and the assumption that there is something in DNA that establishes the unique identity of a nation, can end up being used to provide support for nationalism—what can be described as "genetic nationalism."[23]

Another interesting example comes from Europe. In 1994 a formal collaboration was about to begin between the Centre d'Étude du Polymorphisme Humain (CEPH, Center for the Study of Human Polymorphisms—see Box 2.1 for what polymorphisms are), an important laboratory in France, and Millennium Pharmaceuticals Inc., a US company. In 1993, the CEHP had announced that they were the first to produce a physical map of the human genome, that is, a representation of the distances between particular sites on DNA in terms of units of physical length (Box 1.1). Following this, the collaboration between the CEPH and Millennium was approved by the French government, having as its main aim to discover the genetic basis of particular forms of diabetes. To achieve this, it would be necessary to study the occurrence of the disease in a large set of families whose members had the disease, in order to figure out if there were genes that were commonly found in these people and that were therefore involved in the development of disease. The CEPH had a collection of 800 families (5,500 people); Millennium was well-funded and was developing new and powerful methods for this kind of study. However, in the end the French government blocked the collaboration. The reason was that they wanted to refrain from letting CEPH give away to foreigners something that they considered extremely precious: "French DNA," which was described as "national patrimony."[24]

---

[23] Burton, E. K. (2021). *Genetic crossroads: The Middle East and the science of human heredity.* Stanford, CA: Stanford University Press, pp. 3–9, 260–261.

[24] Rabinow, P. (1999). *French DNA: Trouble in purgatory.* Chicago: University of Chicago Press.

ESSENTIALIZING SOCIAL GROUPS   39

Thinking of Jewish, or Armenian, or French, or any other kind of national DNA might make sense, if one thinks of nations as descent communities. This could be the case under the assumption that these communities are real. However, political scientist and historian Benedict Anderson famously defined the nation as "an imagined political community—and imagined as both inherently limited and sovereign. It is imagined because the members of even the smallest nation will never know most of their fellow-members, meet them, or even hear of them, yet in the minds of each lives the image of their communion." Anderson explained that the nation is imagined as limited because even the largest ones have finite boundaries beyond which other nations lie, however elastic as these boundaries might be. He also explained that the nation is imagined as sovereign because the concept of nation was coined at a time when it was realized that there was a pluralism of religions and so this was the only way for people to peacefully coexist in freedom. So, according to Anderson, people have the feeling of belonging to a limited and sovereign group, and may be willing to sacrifice themselves for it, even though they will never come to know most of its other members—the people they might die for. As he wrote:

> Finally, it [the nation] is imagined as a community, because, regardless of the actual inequality and exploitation that may prevail in each, the nation is always conceived as a deep, horizontal comradeship. Ultimately it is this fraternity that makes it possible, over the past two centuries, for so many millions of people, not so much to kill, as willingly to die for such limited imaginings.[25]

But what made these communities imaginable in the first place? Anderson suggested that it was the interplay between capitalism, print technology, and the decline of linguistic diversity. Before the era of printing, the linguistic diversity in Europe was enormous. With the advent of printing the various spoken languages were assembled into a few "print languages." These provided the basis for the development of national consciousnesses in three distinct ways: (a) by creating unified fields of communication below Latin and above the spoken vernacular languages; (b) by stabilizing languages so that they stopped accumulating changes as they used to do in the past; and (c) by

---

[25] Anderson, B. (1983). *Imagined communities: Reflections on the origin and spread of nationalism.* London: Verso, pp. 6–7.

40 ANCESTRY REIMAGINED

creating languages of different status and power that were different from the older ones. All this, in turn, created the possibility of a new form of imagined community, which provided the foundations for the modern concept of the nation.[26]

Anthropologist Bob Simpson took Anderson's idea further:

> with the rise of the new genetics comes a new vocabulary for grounding difference and similarity as "blood" is displaced by DNA as the essential marker of shared identity and attribute. . . . With this novel means to essentialization comes the possibility of reworking ethnic identities as imagined genetic communities, that is, communities in which the language, concepts and techniques of modern genetic medicine play their part in shaping identity, its boundaries and what is believed to lie beyond.

According to Simpson, DNA technologies give us a new means to create imagined communities that are not political like those described by Anderson, but genetic. Whereas print languages provided in the past a basis for cohesion and belonging, nowadays DNA offers a new basis for these. A common genetic identity, essentialized by reference to a particular, distinctive DNA, can support notions of purity, continuity, and membership in these imagined genetic communities, whether Vikings, Celts, Scots, Greeks, or whatever.[27] Recall from the previous section that Ancestry describes these genetic communities as consisting of people who are connected through DNA. However, these genetic communities are imagined, at least in the same sense that nations are.

I suggest that there is even more to this. It is not only the communities that are imagined but also the categories to which these are supposed to correspond and to which the companies assign the test-taker with their ethnicity estimates. These categories I call *genetic ethnicities*. Further to Anderson's and Simpson's points is that it is not only the community that is imagined but also the essence of the common ethnic identity itself. This is more than just identifying a community that shares a common feature. It is specifying that particular feature, which is materialized and essentialized in the form of DNA. We thus no longer imagine a community that shares a common

---

[26] Anderson, B. (1983). *Imagined communities: Reflections on the origin and spread of nationalism.* London: Verso, pp. 42–46.

[27] Simpson, B. (2000). Imagined genetic communities: Ethnicity and essentialism in the twenty-first century. *Anthropology Today, 16*(3), 3–6.

ESSENTIALIZING SOCIAL GROUPS 41

identity; we imagine a category, and an identity stemming from it, that is based on DNA and that defines the community in an essentialist manner: setting strict boundaries, distinguishing between different types of people, thus prescribing criteria for inclusion and exclusion. This is a strongly essentialist view. A genetic ethnicity is thus defined based on a perceived common genetic identity that is established by reference to specific DNA sequences, common ancestry, and to a well-defined point of origin in time, and sometimes in space.

But genetic ethnicities do not exist in reality—only in our imagination. In fact, the myth that they exist is the one that the present book aims to debunk, as its subtitle indicates. Genetic ethnicities rely on essentialist perceptions of ethnic identity and on the ethnic conception of nations. It is a notion that has been used to support nationalist agendas, explicitly or implicitly, and to support exclusion and inclusion—usually framed in terms of ancestry rather than DNA. Now the problem with DNA ancestry testing is that it unintentionally provides the grounds for the reification of genetic ethnicities. The reason that I argued that this is unintentional is that the respective categories are not constant but change across time for the same company, whereas they also differ from one company to another. Most importantly, these categories usually are distinguished on the basis of geography, not culture, but the respective labels are often the same or similar and are thus easily confused (see Chapter 10). Yet, by receiving test results that assign a person one or more ethnic ancestries, as Karen and Jay did (see the Preface), the implication is that the respective categories exist and are distinguishable based on DNA comparisons.

In subsequent chapters, I consider whether it is possible for human groups that are distinct at the DNA level to exist. But, if national identity depended on DNA, it would be impossible to change. Therefore, if people think national identity cannot change, this might reflect an essentialist view of it. Let us then see whether people indeed think in essentialist terms about national identity.

Essentialism and National Identity

There is a lot of research on race conducted in America (e.g., United States, Canada, Brazil, and elsewhere), and there is a lot of research on ethnicity conducted in Europe. However, with these data it is difficult to make

42 ANCESTRY REIMAGINED

comparisons among countries. The reason for this is not only that race and ethnicity can be conceptualized differently in different regions of the world (see Chapter 1), but also that they can be confused. In contrast, nationality is relatively easier to define based on one's citizenship and official documents. In order to obtain a general picture of how people think about nationalities, it is useful to consider cross-national surveys. One such survey is the 2018 IPSOS study, titled *The Inclusiveness of Nationalities: A Global Advisor Survey*.[28] Let us consider one of its questions (I have numbered the items for convenience). The participants were asked whether a person who was raised and born in their country, but who came from one of nine different continental regions, would become a true national (items 2–10); as well as whether a person of their own nationality would maintain it if they lived abroad (item 1).

Q3. For each item in the list below, please indicate if you think a person like this is or is not a real [country demonym]
1. Someone born and raised abroad by [country demonym] parents
2. Someone born and raised in [country] whose parents immigrated from Europe or North America
3. Someone born and raised in [country] whose parents immigrated from Latin America
4. Someone born and raised in [country] whose parents immigrated from the Caribbean
5. Someone born and raised in [country] whose parents immigrated from any part of Africa excluding North Africa
6. Someone born and raised in [country] whose parents immigrated from North Africa or the Middle East (e.g., Algeria, Egypt, Iraq, Saudi Arabia)
7. Someone born and raised in [country] whose parents immigrated from South Asia (e.g., India, Pakistan, Bangladesh)
8. Someone born and raised in [country] whose parents immigrated from East Asia (e.g., China, Korea, Japan)
9. Someone born and raised in [country] whose parents immigrated from Southeast Asia (e.g., Vietnam, Indonesia, Philippines)
10. Someone born and raised in [country] whose parents immigrated from a Pacific Island (e.g., Papua New Guinea, Fiji, Samoa)

---

[28] Boyon, N. (2018). The inclusiveness of nationalities: A global advisor survey. *Ipsos Public Affairs*, 2018–06. The survey was conducted between April 20 and May 4, 2018, in 27 countries with 20,767 adults.

ESSENTIALIZING SOCIAL GROUPS   43

Based on participants' responses, the researchers calculated a net score per country. This was the percentage of participants who said that the child of immigrants born and raised, say, in Greece is a Greek, minus the percentage of participants who said that the child of immigrants born and raised in Greece is not a Greek—in both cases for a particular continental region of origin. Therefore, a positive score indicates that more people in that country agreed rather than disagreed that the child of immigrants born and raised in their country would become a true national; whereas a negative score means the opposite. This entails that in all countries, both those with positive scores and those with negative scores, there were both people who agreed and people who disagreed with the statements. So, when I write below that some countries are more or less essentialist about nationalities than others, this is a vague description based on averages; there is interesting variation within each country.

One interesting finding was that in most countries people thought that the child of immigrants born and raised in their country was, or was not, a true national, more or less in the same manner for all nine continental regions of origin. This means that the region from which the immigrant parents came did not seem to make a very big difference, even though there was some interesting variation. Japan and China had negative scores only. This indicates that most people in these countries thought that people from any continental region would not become a true national if they were raised and born in Japan and China. Saudi Arabia, Malaysia, Serbia, and Turkey had negative scores for all continental regions except for one in each case, which indicates that participants thought of people coming from that particular region as being closer to them than those coming from the other eight regions. For Turkey and Serbia that particular region was Europe/North America; for Malaysia it was South Asia; whereas for Saudi Arabia it was Middle East/South Africa. Therefore, the only continental region that got an overall positive score in all these cases was one that is close geographically or culturally to the country of participants. Some other interesting variation was found in Italy, where the lowest positive score, quite lower than those for the other regions, was given to people coming from the Pacific, whereas Middle East/Northern Africa got the lowest score in India and Poland. Overall, where immigrants came from did not matter a lot, even though countries differed in how people perceived the possibility of integration of immigrants.

The most interesting finding, however, is shown in Figure 2.1. It presents the average score for each country for items 2–10 (horizontal axis). The countries that have the highest scores (those toward the rightmost part of

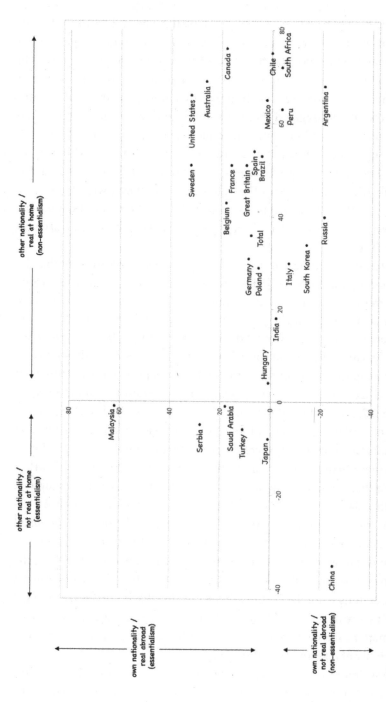

Figure 2.1 Distribution of countries depending on how essentialist or nonessentialist their scores were about their own or other nationalities. (Based on data from Boyon, N. (2018). The Inclusiveness of Nationalities: A Global Advisor Survey. *Ipsos Public Affairs*, 2018–06, pp. 36–37.)

ESSENTIALIZING SOCIAL GROUPS 45

the figure) are those in which more people considered that a person born and raised in their country was a true national rather than not. Therefore, we can assume that the people in the countries with the highest scores for these items were overall less essentialist about nationality, as they seemed to consider more important for a person's nationality the place where that person was born and raised, rather than the place where that person's parents came from. In contrast, the countries with the lowest scores, especially the negative ones (toward the leftmost part of the figure) were more essentialist about nationality (particularly Japan, China, Turkey, Saudi Arabia, Serbia, and Malaysia, as already discussed). Figure 2.1 also presents the score for item 1 for each country (vertical axis). The countries that have the highest scores for this item are those in which more people considered that one of them living abroad would maintain their nationality rather than not. Therefore, these were also overall more essentialist about nationality. Let us look at these findings in more detail.

Take a look at the upper left quadrant in Figure 2.1. Especially Malaysia, but also Serbia, Saudi Arabia, and Turkey, had a negative score for items 2–10, which means that more people considered that the child of immigrants born and raised in their country would not become a true national rather than not; they also had positive scores for item 1, which means that more people considered that other people of the same nationality with them would maintain their nationality abroad rather than not. Therefore, more people in these countries thought in essentialist terms about nationality: they considered that people would maintain their nationality in another country, but thought so mostly about their own people living abroad than about foreigners living in their own country. Now take a look at the lower right quadrant. People in Argentina, Russia, South Korea, and Italy had a positive score for items 2–10, which means that more people considered that the child of immigrants born and raised in their country would become true nationals rather than not; they also had negative scores for item 1, which means that more people considered that people of the same nationality with them would not maintain their nationality abroad rather than not. Therefore, more people in these countries thought in nonessentialist terms about nationality.

Whether essentialist or nonessentialist about nationality, people in all the aforementioned countries were overall consistent in their thinking. But there were also countries in which people were overall inconsistent. For instance, people in China had negative scores for all items. This means that many people in China believed that the child of immigrants born and raised in their country would not become a true national (essentialism about nationality), as well as that one of them would not maintain their own nationality abroad (nonessentialism about nationality). For many people in Sweden, the United States, Australia, Canada, France, and Belgium, the situation was exactly the opposite: for them a foreigner born and raised in their country would become a true national (nonessentialism about nationality), whereas one of them would maintain their own nationality abroad (essentialism about nationality).

The important conclusion from these findings is that we cannot easily generalize about whether people think in essentialist terms about nationalities. The attitudes are different in different countries, and there are also different ways of thinking about the members of one's ingroup compared to the members of outgroups. It is worth pointing out that, overall, there were more countries where people were on average less essentialist about nationality (especially about the nationality of others but also about their own in some cases) than countries where people were on average more essentialist about it. Argentina was the less essentialist country of all on both grounds, whereas Malaysia and China were the most essentialist ones, the former about their own nationality and the latter about that of others.

These results are perhaps not surprising. People nowadays are aware that nationality is acquired at birth, but it can also change during one's life. For instance, a person may emigrate and eventually acquire the nationality of the host country. In other cases, people can have two or more nationalities, for instance, if their parents come from different countries. In some cases, nationality can be acquired just because a person happened to be born in a country (what is described as birthright citizenship). Therefore, and in contrast to ethnicity and race, nationality is not necessarily inherited by one's parents and therefore is less restrictive. Given all this, it is not surprising that more people in most of the countries considered in this study thought that nationality can change. Race and ethnicity are different in this respect, and they are considered in detail in Chapter 3.

But what do people see as the essential feature of nationality? This was investigated by another survey for the PEW Research Center that drew on data from national samples in 14 countries.[29] Participants were asked

---

[29] Pew Research Center, February, 2017, "What It Takes to Truly Be 'One of Us'" (https://www.pewresearch.org/global/2017/02/01/what-it-takes-to-truly-be-one-of-us/; accessed January 20, 2022). This survey involved 14,514 participants in total, and it took place between April 4 and May 29, 2016, in the United States, Canada, France, Germany, Greece, Hungary, Italy, Netherlands, Poland, Spain, Sweden, United Kingdom, Australia, and Japan).

ESSENTIALIZING SOCIAL GROUPS   47

how important they considered each of the following for being a true national: (a) having been born in the country; (b) speaking the national language; (c) subscribing to the dominant religion; and (d) sharing the country's customs and traditions. Let me point out that none of these questions refers to ancestry, or the ethnicity of one's ancestors. All four questions rather refer to one's birthplace or to specific cultural features. Speaking the language of the country was considered as the most important feature for being a true national by the majorities in most countries. This ranged from 59% in Canada and Italy to 84% in the Netherlands and 81% in Hungary and the United Kingdom. The next most important feature was sharing national customs and traditions—perhaps less in Sweden (26%) and Germany (29%), but quite a lot in Greece (66%) and Hungary (68%). Less than half of the people in most countries thought it was very important for a person to be born in their country in order to be considered a true national. However, this was considered as very important by half of the participants in some countries, such as Hungary, Greece, and Japan. Finally, the survey asked about the importance of religion for national identification (a question not asked in Japan). There was quite a variation among countries, from 7% in Sweden to 54% in Greece.

Why was speaking the language of the country considered the most important feature of nationality? Academic and politician Michael Ignatieff has argued that belonging means being recognized and being understood. "To belong is to understand the tacit codes of the people you live with; it is to know that you will be understood without having to explain yourself. People, in short 'speak your language.'" Ignatieff also argued that this is why the protection and conservation of a nation's language is a deeply emotional issue: it provides the essential form of belonging, being understood, which is more important than homeland and history.[30] Well, this is an argument that might make sense to many people. It certainly makes sense to me, as I grew up in Greece. In the aforementioned PEW study, 76% of people from Greece stated that being able to speak the national language is very important for being a true national. However, for the past 10 years I have been living in Switzerland, and I have been experiencing something quite different, perhaps entirely the opposite. In Switzerland, the French-speaking, German-speaking, Italian-speaking, and Romansh-speaking communities, which might even be considered as different ethnic groups, are all united under the Swiss national identity. Indeed, Switzerland is

---

[30] Ignatieff, M. (1994). *Blood and belonging: Journeys into the new nationalism.* London: Vintage, p. 7.

48 ANCESTRY REIMAGINED

an exemplar case of a multiethnic nation, defined as comprising several ethnic (or, more precisely, ethno-linguistic) communities.[31] I do not know if exceptions confirm the rule or question it, but I believe that it is useful to consider them.

Keeping in mind that the two surveys presented in this section did not include people from any African country, we can conclude that there is an interesting variation among countries as to whether people think of nationality in essentialist terms, with language being its most important feature. Beyond this, in the current era of globalization and global migration, it is possible for a country to receive immigrants from various places in the world. Based on the surveys considered here, we can arrive at the vague generalization that in many countries in the world (Figure 2.1, right) immigrants can be considered as true nationals of a country if they are born and raised there, and also speak the language. This notwithstanding, the results of DNA ancestry tests are framed in terms of racial or ethnic groups. This is done either with reference to a continent, which is perceived to correspond to race (and by some to biological race—see Chapter 7), or with reference to a subcontinental region, which is perceived to correspond to ethnicity.

Let us then consider race and ethnicity in some detail.

---

[31] Wimmer, A. (2011). A Swiss anomaly? A relational account of national boundary-making. *Nations and Nationalism, 17*(4), 718–737.

3

From Race to Ethnicity in Ancestry Testing

Race and Science

Our species is *Homo sapiens*; the first word (*Homo*) denotes the genus, and the two words together denote the species. This binomial nomenclature for classifying organisms was established by the Swedish naturalist Carl Linnaeus in the mid-17th century, and it is still used today. We currently know that several other human, or human-like, species have lived on Earth, including *Homo habilis*, *Homo rudolfensis*, *Homo ergaster*, *Homo erectus*, *Homo heidelbergensis*, *Homo neanderthalensis*, and *Homo floresiensis*. Most, but not all, of these seem to have become extinct before modern humans, *Homo sapiens*, evolved and dispersed all over the earth. These conclusions are based on two kinds of biological data: fossils of human skeletons, in which organic material has been replaced by the material from the rock in which it is found, and DNA either from contemporary populations (modern DNA) or from the preserved bone fragments of ancient individuals (ancient DNA). Even though the current picture as it emerges from the fossils that have been found is rather fragmented and we do not know the exact relations among the various species, it suffices to establish the considerable diversity in our lineage during the last few hundred thousand years of human evolution. We now know that our ancestors coexisted for some time with the Neanderthals (*Homo neanderthalensis*) and the so-called Denisovans, another extinct species or subspecies of *Homo*.[1] Despite any variation in body size, shape, and skin color, all humans living today clearly belong to a single species, *Homo sapiens*, which is characterized by shared features such as a narrow pelvis, a large brain within a globular braincase, and reduced size of the teeth and surrounding skeletal architecture.

---

[1] For an overview of what we know about human evolution, see Wood, B. (2019). *Human evolution: A very short introduction* (2nd ed.). Oxford: Oxford University Press; Tattersall, I. (2022). *Understanding human evolution*. Cambridge: Cambridge University Press.

Ancestry Reimagined. Kostas Kampourakis, Oxford University Press. © Oxford University Press 2023.
DOI: 10.1093/oso/9780197656341.003.0003

50 ANCESTRY REIMAGINED

In the first nine editions of his book *Systema naturae*, published between 1735 and 1756, Linnaeus identified four different varieties of humans: *Homo europaeus albus* (European white), *Homo americanus rubescens* (American reddish), *Homo asiaticus fuscus* (Asian tawny), and *Homo africanus niger* (African black). This division corresponded to the then known four continents of the world. In the tenth edition of his book, published in 1758, Linnaeus changed the hierarchy and the names to *Homo sapiens americanus, Homo sapiens europaeus, Homo sapiens asiaticus*, and *Homo sapiens afer* (there was a first variety of wild children and youngsters described as "Ferus" and a last one of "Monstrous" humans, which included groups that were allegedly shaped by their environment). Linnaeus also added the four temperaments to these four varieties of humans. He thus described *Americanus* as red, choleric, and straight; *Europaeus* as white, sanguine, and muscular; *Asiaticus* as sallow, melancholic, and stiff; and *Africanus* as black, phlegmatic, and lazy. Even though Linnaeus revised this hierarchy several times, his description of *Africanus* was always at the bottom of the list, being the longest, the most detailed, and also the most negative.[2] This kind of thinking was hardly unusual among Europeans at the time.[3] Various naturalists thought that the anatomical features of non-Europeans were underdeveloped, and that in some cases those people were closer to apes than to other humans. They thus developed a hierarchy in which Europeans were on the top, whereas Africans and others were ranked lower.

Philosopher Kwame Anthony Appiah has argued that there are three important views about race, which emerged during the 19th century:

- *Racial fixation.* This is the idea that the characteristics of individuals could be explained as a product of their race.
- *Typological thinking.* A consequence of racial fixation is that each individual expresses the shared nature of their race, and so they are representative of their "type."

---

[2] https://www.linnean.org/learning/who-was-linnaeus/linnaeus-and-race (accessed January 19, 2022).

[3] Even though this division can be seen to imply distinct racial groups, Linnaeus himself did not present these divisions of *Homo* as discrete and stable types; in other words, he did not essentialize them. Rather, what he was interested in was the division based on geography. See Müller-Wille, S. (2015). *Linnaeus and the four corners of the world.* In K. A. Coles, R. Bauer, Z. Nunes, and C. L. Peterson (Eds.), *The cultural politics of blood, 1500–1900* (pp. 191–209). Basingstoke: Palgrave MacMillan.

FROM RACE TO ETHNICITY IN ANCESTRY TESTING   51

- *Essentialism*. Behind both racial fixation and typological thinking lies the idea that each individual has something derived from their race that explains their mental and biological potential. This racial essence is transmitted across generations, from parents to offspring.[4]

Reversing this order, the idea is this: there are inherent racial essences that are expressed in each one of us and that explain not only how we look but also who we are and what we do. This is where biology comes in: as explained in Chapter 2, DNA serves as a placeholder for these racial essences. This entails that the members of the same racial group should have similar DNA with one another, which at the same time should be different from the DNA of members of other racial groups.

Let us apply to race the set of interrelated beliefs that essentialist thinking consists of, which we considered in Chapter 2:[5]

I. Particular categories distinguish between fundamentally different kinds of people → Black and White people are fundamentally different.
II. The boundaries that separate these categories are strict and absolute, so that a person who belongs to a particular category cannot belong to another → A person can be either Black or White.
III. These categories are homogeneous; that is, their members share fundamental similarities with one another and have fundamental differences from members of other groups → White people are more similar to other White people in terms of their skin color than to Black people.
IV. All this is due to internal factors that make the members of each category what they are, which is the category essence → Black and White people have different DNA sequences related to their skin color.

Therefore, according to an essentialist view of race, races such as Black and White are fundamentally different, have strict boundaries, and are internally homogeneous, with all these features being due to their differences in DNA,

---

[4] Appiah, K. A. (2018). *The lies that bind: Rethinking identity*. London: Profile Books, pp. 112–117.
[5] Rhodes, M., and Kelsey, M. (2020). What is social essentialism and how does it develop? In M. Rhodes (Ed.), *The development of social essentialism* (pp. 1–30). San Diego: Academic Press.

52 ANCESTRY REIMAGINED

and this constitutes their essence. In other words, the essentialist view of race is the view of biological race.

Scientific writings of the mid-19th century provide characteristic accounts of this view. For instance, consider the following excerpt from a book by physician (and slave-owner) Josiah Clark Nott:

> But the chief value of these osteological differentia lies in their perfect applicability to man, and the facility with which they enable us to distinguish between the various human types. Thus, in the best developed and most intellectual races, the supra-orbital ridge is smooth, well carved, and not much developed; as we descend towards the lower types, it becomes more and more marked, until, in the African and Australian heads, it has attained its maximum development. In the Orang, this feature begins to assume a greater importance, while in the Chimpanzee, its enormous size renders it a characteristic mark. Here, then, is the evidence, to some extent, of gradation, in a seemingly exclusive ethnographic mark, whose significance is elucidated by a resort to anthropology.[6]

In short, there is a gradation of the supra-orbital ridge (also called the brow ridge, a bony ridge located above the eye sockets of all primates) that starts from the most marked versions in chimpanzees and orangutans, and then goes through the lower human types, Africans and Australians, to reach "the best developed and most intellectual races." Can you guess who the latter are? The Europeans, of course. On the same page with this text there is a plate, shown here in Figure 3.1. It shows the skull of a "Negro," an "Australian," and a "Greek." The first two are described on the previous page as "the adult or permanent forms of the lower types ... of men." The "Greek" skull is described as "a well-known representation of the highest form of the "human head divine"; in other words, the ideal (or perhaps idealist) representation of the human skull. Nott noted that the greatest resemblances between humans and apes were "to be sought for in or between the lower types of each, and not between the lowest man and highest monkey, as is generally supposed,"[7] which can be interpreted as that he did

---

[6] Nott, J. C. (1857). *Indigenous races of the earth, or, New chapters of ethnological inquiry: Including monographs on special departments of philology, iconography, cranioscopy, palaeontology, pathology, archaeology, comparative geography, and natural history.* Philadelphia: JB Lippincott & Company, p. 207 (available at https://collections.nlm.nih.gov/catalog/nlm:nlmuid-60420070R-bk).

[7] Nott, J. C. (1857). *Indigenous races of the earth, or, New chapters of ethnological inquiry: Including monographs on special departments of philology, iconography, cranioscopy, palaeontology, pathology,*

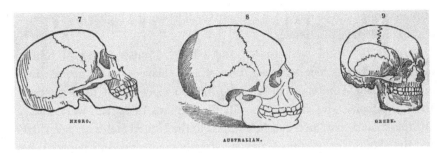

Figure 3.1 The skulls of a Negro and an Australian compared to the idealized skull of a Greek in an 1857 book by Josiah Clark Nott: Nott, J. C. (1857). *Indigenous races of the earth, or, New chapters of ethnological inquiry: including monographs on special departments of philology, icongraphy, cranioscopy, palaeontology, pathology, archaeology, comparative geography, and natural history.* Philadelphia: JB Lippincott & Company, p. 207 (book considered to be in the public domain https://collections.nlm.nih.gov/catalog/nlm:nlmuid-60420070R-bk).

not consider the lower human forms, Africans and Australians, as being closer to apes than to other humans. Still, they were considered as being "lower." But why?

Long story short, during the era of European colonialism (roughly between the 15th and the 19th century), European explorers encountered humans they had never seen before, who differed in various biological characteristics from the stereotypical White European. Because these previously unseen people were less advanced technologically than Europeans, they were easy to conquer and exploit—this often turned them into slaves, as the exploitation of newly conquered lands required workers. Of course, exploiting other humans as slaves was unethical for the mostly Christian European colonists. But this problem was solved for many when some naturalists argued that non-Europeans were of lower human status, and in a sense less human, than Europeans. This was eventually what brought the idea of distinct human biological races to the fore, and unfortunately 19th-century science lent a hand to that—this is why the view is sometimes described as scientific racism. As anthropologist Jonathan Marks has explained, this was the outcome of two related fallacies: *racialism*, the false idea that the human species can be naturally divided into distinct groups; and *racism*, the morally corrupt idea that

archaeology, comparative geography, and natural history. Philadelphia: JB Lippincott & Company, p. 205.

54 ANCESTRY REIMAGINED

human groups can be ranked in a hierarchy of lower and higher ones because of their biological differences.[8]

Perhaps the most well-studied and widely discussed case of human racist exploitation has been the Transatlantic Slave Trade, due to which many African Americans today have limited clues about their ancestry. Their ancestors, who came from different regions of Africa, were sold to Europeans and were forced to be shipped to the United States between 1619 and the 1800s. Henry Louis Gates, Jr., already mentioned in Chapter 1, has been the host of several TV shows about ancestry, including the one called *African American Lives.*[9] Gates was interested in the family histories and the genealogies of other African Americans, and he thus invited several well-known ones as guests to his show. These included poet Maya Angelou, actor Morgan Freeman, television host Oprah Winfrey, and musician Quincy Jones. According to Gates, most African Americans living today descend from the approximately 455,000 Africans who were transferred from Africa to the United States (many more, around 12 million, Africans were taken to the Caribbean and South America). Those who managed to survive the trip, "the dreadful Middle Passage," Gates continued "endured a lifetime of unimaginable hardship, bound by people who carefully and willfully did all they could do in every possible way to strip away every aspect of their slaves' humanity."[10] There were around 4 million enslaved persons in the United States in the middle of the 19th century, most of whom were freed in 1865, at the end of the Civil War. By that time, most of them had been born into slavery and had limited clues about their regions of origin in Africa. This is why genealogy and ancestry discussions today are much framed in terms of race. As sociologist Alondra Nelson has shown, DNA tests related to ancestry have been invoked in various ways in order to deal with the consequences of the slave trade. These ways include fostering reconciliation, establishing ties with the homelands in Africa, reconsidering citizenship, and even claiming legal reparations for harms caused due to slavery.[11]

If 19th-century science provided scientific evidence for a biological conceptualization of race, it was 20th-century science with eugenic policies,

---

[8] Marks, J. (2017). *Is science racist?* Cambridge: Polity; see also Graves, J. L., and Goodman, A. H. (2021). *Racism not race: Answers to frequently asked questions.* New York: Columbia University Press.

[9] https://www.imdb.com/title/tt0773257/ (accessed January 19, 2022).

[10] Gates Jr, H. L. (2017). *In search of our roots: How 19 extraordinary African Americans reclaimed their past.* New York: Skyhorse Publishing, p. 15.

[11] Nelson, A. (2016). *The social life of DNA: Race, reparations, and reconciliation after the genome.* Boston: Beacon Press.

resulting in the sterilization of numerous people in the United States and some European countries,[12] and eventually racist discrimination leading to the Holocaust that took the inhumanity of slavery to an unprecedented level. Of course, there were reactions early on. Among others, anthropologist Franz Boas, cultural anthropologist Alfred Kroeber, physical anthropologist Sherwood Washburn, and geneticist Theodosius Dobzhansky attempted to discredit scientific racism. The scientific advancements of the 1930s and the 1940s (what came to be called the Modern Evolutionary Synthesis) rejected the typological thinking that was a prerequisite for racist and eugenic theorizing. Rather than being considered as types, races were described as "biogeographically distinctive populations," which contained a lot of genetic diversity but which could also interbreed with one another.[13] Still, others remain unconvinced.

A proposal that resulted in great controversy was made in 1962 by anthropologist Carleton Coon. He suggested that five lineages of *Homo erectus* had been evolving independently and in parallel, two in Africa, one in Europe, one in Asia, and one in Australia. These independently evolving lineages eventually gave rise to what Coon described as the modern races of *Homo sapiens*: "Capoid," "Congoid," "Caucasoid," "Mongoloid," and "Australoid." Coon also used the term "Negroid" to "denote a condition, not a geographical subspecies." As he put it:

> If all races had a recent common origin, how does it happen that some peoples, like the Tasmanians and many of the Australian aborigines, were still living during the nineteenth century in a manner comparable to that of Europeans of over 100,000 years ago? Either the common ancestors of the Tasmanians cum Australians and of the Europeans parted company in remote Pleistocene antiquity, or else the Australians and Tasmanians have done some rapid cultural backsliding, which archaeological evidence disproves.[14]

I think that the consequences that such a view can have for mistakenly justifying racial discrimination are obvious. As a result, during the 20th

[12] Kevles, D. J. (1995). *In the name of eugenics: Genetics and the uses of human heredity.* Cambridge, MA: Harvard University Press.

[13] Jackson, J. P., and Depew, D. J. (2017). *Darwinism, democracy, and race: American anthropology and evolutionary biology in the twentieth century.* London: Routledge.

[14] Coon, C. S. (1962). *The origin of races.* New York: Alfred Knopf, pp. 4–5 (available at https://arch ive.org/details/B-001-001-289).

56 ANCESTRY REIMAGINED

century, many scientists ended up not being happy with the use of term "race" because of its political connotations that pertained to discrimination and inequalities. However, this does not mean that they thought that races did not actually exist at the biological level.[15]

For instance, as late as 1970, Dobzhansky wrote that whereas races were in general allopatric (meaning that they lived in different areas), humans were an exception: "Civilization created a variety of social forces that make possible, at least for a time, the sympatric coexistence of human races" ("sympatric" meaning living in the same area). He also noted that a race "consists of individuals who differ genetically among themselves. It is important to realize that similar genetic elements are involved in individual and in race differences. . . . Blue-eyed individuals are not a race distinct from brown-eyed ones, yet eye color is one of the trait distinguishing races." Dobzhansky thus rejected the idea that the members of each race were essentially similar to one another, and essentially different from the members of other races. However, he wrote: "The obvious fact is, however, that members of the same species who inhabit different parts of the world are often visibly and genetically different. This, in the simplest terms possible, is what race is as a biological phenomenon."[16] Dobzhansky thus rejected the idea that humans are divided in superior and inferior groups, but he did accept that they are divided into biologically distinct, although interbreeding, groups nevertheless.

In 2001 and 2002, sociologist and demographer Ann Morning conducted interviews with 22 academics in biology and 19 academics in anthropology in the United States. These academics were found to hold a variety of views about the nature of race.[17] In particular, 16 of these academics (6 anthropologists and 10 biologists) described race as a biological phenomenon. Most of them explained race as the outcome of evolutionary processes in (more or less) isolated human populations that did not interbreed. Because of these, each race has distinct genetic characteristics that have evolved through independent processes of adaptation to local environments. This notwithstanding, half of these academics pointed out the limitations of racial classification schemes, as they perceived the boundaries among the various races not to be strict. The other 25 academics (12 biologists and 13

---

[15] For a detailed and informative account, see Reardon, J. (2005). *Race to the finish: Identity and governance in an age of genomics.* Princeton, NJ: Princeton University Press.

[16] Dobzhansky, T. (1970). *Genetics of the evolutionary process.* New York: Columbia University Press, pp. 267–269.

[17] Morning herself pointed out the small size and the particular features of the sample (66% male, 83% White, and 90% left-leaning in their political orientation).

anthropologists) all rejected a biological concept of race, expressing two different views: the "anti-essentialist" view, which considered the idea of race as a biological phenomenon to be false, with reference to evidence about the genetic similarity among humans; and the "constructivist" view, according to which race was perceived as a way of thinking about human difference that was socially constructed with reference to historical, social, or political factors. Most accepted one of these views, but some accepted both of them.[18]

This results in what seems like an interesting continuum of views. But what exactly does it mean that race is socially constructed?

Race as a Social Construct

Many professional associations nowadays reject biological conceptions of race. For instance, The American Association of Biological Anthropologists notes: "The distribution of biological variation in our species demonstrates that our socially-recognized races are not biological categories. . . . The racial groups we recognize in the West have been socially, politically, and legally constructed over the last five centuries."[19] The American Society of Human Genetics "affirms the biological reality that we are one people, one species, and one humanity. . . . Genetics demonstrates that humans cannot be divided into biologically distinct subcategories or races, and any efforts to claim the superiority of humans based on any genetic ancestry have no scientific evidence."[20] These are a constructivist and an anti-essentialist stance, respectively (see the previous section). Another interesting example is the Jena declaration on race, published in 2019 by the Friedrich Schiller University in Jena, Germany. Because that university supported racial studies during the Nazi era, its faculty felt that they had a responsibility to address the issue of

---

[18] Interestingly, scholars of the same discipline were divided in their views: 32% of anthropologists and 46% of biologists thought in essentialist terms about race, whereas 47% of anthropologists were constructivists about race and 54% of biologists were either constructivists or anti-essentialists. Whereas one might have expected biologists to be more essentialist and less constructivist about race than anthropologists, it seems that they can have a variety of views about race. See Morning, A. J. (2011). *The nature of race: How scientists think and teach about human Ddfference.* Berkeley: University of California Press, Chapter 4; Morning, A. (2018). The constructivist concept of race. In K. Suzuki and D. A. von Vacano (Eds.), *Reconsidering race: Social science perspectives on racial categories in the age of genomics* (pp. 50–61). New York: Oxford University Press.

[19] https://physanth.org/about/position-statements/aapa-statement-race-and-racism-2019/ (accessed January 19, 2022).

[20] https://www.ashg.org/publications-news/ashg-news/statement-regarding-good-genes-human-genetics/ (accessed January 19, 2022).

58 ANCESTRY REIMAGINED

Box 3.1 How Is Skin Color Formed?

Skin consists of the epidermis, a thin outer layer that contains pigment-producing and immune cells, and the dermis, a thick inner layer that contains blood vessels, sweat glands, sensory receptors, and hair follicles. Skin color depends on several of these substances, but the most important one is melanin, a pigment produced in particular skin cells called melanocytes, which are found at the lowest level of the epidermis and right above the dermis. Melanin is made within small cellular vesicles called melanosomes. These are then transferred to the adjacent skin cells, where they aggregate above their nucleus where the cell DNA is found and absorb ultraviolet (UV) radiation from the sun. This is a process that occurs in all people. The different colors that we see are mainly due to the different amounts and kinds of melanin that the melanocytes of each one of us produce, as well as due to the different sizes and distribution of melanosomes in our skin. This is why what actually exists is a continuum of skin colors.

Jablonski, N. G. (2012). *Living color: The biological and social meaning of skin color*. Berkeley: University of California Press, pp. 9–14.

---

defining human races based on biological data. Their declaration concludes by highlighting the responsibility of scientists: "So, let us ensure that people are never again discriminated against on specious biological grounds and remind ourselves and others that it is racism that has created races and that zoology/anthropology has played an inglorious part in producing supposedly biological justifications. Today and in the future, not using the term race should be part of scientific decency."[21]

But all this results in a paradox. The distinction among different races is often based on a person's skin color. This is unquestionably a biological characteristic, which depends on particular genes and emerges during development (Box 3.1).[22] This is probably why people and scientists, historically and today, have argued that races are biological. But if skin color is a biological

---

[21] https://www.uni-jena.de/en/190910-jenaererklaerung-en (accessed January 20, 2022).

[22] Simply put, genes are DNA sequences that encode the information for the synthesis of functional molecules, such as melanin. In doing so, genes are implicated in the development of traits such as skin color. See Kampourakis, K. (2021). *Understanding genes*. Cambridge: Cambridge University Press. For the genes associated with skin color, see Rocha, J. (2020). The evolutionary history of human skin pigmentation. *Journal of Molecular Evolution*, 88(1), 77–87).

characteristic that distinguishes among races, why then are races not biological? This is due to the confusion of continental divisions and racial divisions, which results in the identification of racial groups with particular continents. It is, of course, true that particular features are found more often in particular regions of the world rather than others. For instance, dark skin coloration is found more often in people living in Africa; the epicanthic fold, a skin fold of the upper eyelid that covers the inner corner of the eye, is found more frequently in people living in Eastern Asia. But this does not mean that only Africans have dark skin color, or that only Eastern Asians have the epicanthic fold. Nor does this mean that all Africans have the same dark skin color, or that all Eastern Asians have the same epicanthic fold. These are stereotypes that blind us to the enormous variation that exists among people regarding these features. Because of this variation, we cannot say that these characteristics are exclusive, and therefore distinctive, of particular races. This is why the geographical distribution of a few biological features is not a good criterion for classification. This does not deny the fact that particular biological features such as skin pigmentation or physiognomy can be used in order to distinguish one from the other.[23] A person with dark skin color is likely to be African, but dark skin color is also found in people who live in Asia (for instance, India). The fact that dark skin color is found more often in Africans than in other groups does not mean that there is an underlying racial, biological essence that can account for that.

A key to resolving the paradox is that any differences in skin color (and in any other biological characteristic for that matter) among humans are differences in degree rather than kind. As anyone can observe in a cosmopolitan city like New York or London, there are many intermediate conditions between being Black or White. That the diversity of human skin color is enormous has been excellently and elegantly illustrated by artist Angélica Dass in her photographic project called *Humanæ*. This is a project that has aimed to document the actual colors of humanity in the place of the stereotypical labels of "white," "red," "black," and "yellow," usually associated with race. Dass has taken more than 4,000 images in 36 cities, from 20 different countries all over the world.[24] In each case, she has colored the background of each image with

---

[23] Rutherford, A. (2020). *How to argue with a racist: History, science, race and reality*. London: Weidenfeld and Nicolson, p. 21.

[24] Arteixo, Madrid, Barcelona, Getxo, Bilbao, and Valencia (Spain); Paris (France); Bergen (Norway); Winterthur and Chiasso (Switzerland); Groningen and The Hague (Netherlands); Dublin (Ireland); London (United Kingdom); Tyumen (Russia); Gibellina and Vita (Italy); Vancouver and Montreal (Canada); New York, San Francisco, Gambier, Pittsburgh, and Chicago (United States);

60 ANCESTRY REIMAGINED

a color tone identical to a sample of 11 × 11 pixels taken from the nose of the individual and matched with the industrial pallet Pantone®. The outcome is stunning, as it shows the enormous diversity in skin colors that actually exists. As noted on the project website, the outcome "in its neutrality calls into question the contradictions and stereotypes related to the race issue."[25] Therefore, what actually exists is not a strict division between people with different skin colors (e.g., either Black or White), but rather a continuum of skin colors. Of course, we do distinguish among people based on their skin color, but there is no absolute or strict criterion for doing so; it is mostly relative. Furthermore, apparently similar skin colors can have a very different genetic and developmental basis (Box 3.2). This is why it is mistaken to think about skin color in terms of racial essences. If the genetic and developmental basis of similar skin colors can be very different, then races that are distinguished on the basis of skin color cannot be biological categories. In such a case, races can only be social constructs.[26]

Let us clarify when something is a social construct. All our knowledge takes the form of mental representations. Think of a Black person. It is likely that my own and your own representation of a Black person are different, depending on our social experiences. I might think of NBA star Lebron James as an example, whereas you might think of the famous singer Beyonce. They both have specific features, such as skin color for instance, on which we could base their classification as Black. However, it is not just our representations of "Black" that are socially constructed based on our experiences; rather, it is the category "Black" itself. The key point is that whereas our representation of a Black person is constrained by what we know as a fact about this group, such as—again—their skin color, it is this very representation of "Black" that can be invoked to explain the features of people such as Lebron James

---

Quito (Ecuador); Valparaíso (Chile); Sao Paulo and Rio de Janeiro (Brazil); Córdoba (Argentina); New Delhi (India); Daegu (South Korea); Wenzhou and Shanghai (China); Ciudad de México and Oaxaca (Mexico); and Addis Abeba (Ethiopia).

[25] https://angelicadass.com/photography/humanae/ (accessed April 11, 2022). I am grateful to Stefan Burmeister for bringing this project to my attention.

[26] This is my own view, of course. From a philosophical point of view, there can exist different arguments about race: (1) that races are first and foremost sociopolitical groups; (2) that races are not only political but also, and importantly, cultural groups; (3) that racial groups can be distinguished from one another by biological features, which nevertheless do not justify any biological hierarchy; or (4) that there exist groups that have visible features that are held disproportionally, which entail that either race is not real, or if it is real, it is in a way that is neither social nor biological. See Glasgow, J., Haslanger, S., Jeffers, C., and Spencer, Q. (2019). *What is race? Four philosophical views.* Oxford: Oxford University Press.

Box 3.2 How Have Different Skin Colors Evolved?

Our genus, *Homo*, initially evolved in Africa under high ultraviolet (UV) radiation, and so individuals with enhanced melanin pigmentation, that is dark skin color, had a survival advantage as it provided protection against folate deficiency, as well as against the damage of their DNA by UV radiation. On the one hand, UV radiation can cause folate breakdown in the skin, which is problematic as folate is necessary for many biochemical processes, including melanin production. On the other hand, UV radiation can cause mutations in DNA that might in turn contribute to skin cancer. At the same time, skin color was not too dark so as not to allow vitamin D production. Vitamin D has many important roles in our body and is produced from cholesterol thanks to UV radiation. So survival (and subsequently reproduction) is more likely when skin pigmentation allows for an amount of UV radiation that is both sufficient to produce vitamin D and does not damage folate or DNA. Among these, protection of folate seems to be the most important factor for the evolution of dark skin color.[1] What is most interesting and important is that what is apparently the same skin color can be due to different underlying developmental processes, as there are at least 20 genes associated with skin color variation. The reason that similar skin colors can be due to different developmental processes and genes lies in our evolutionary history: humans have evolved to have similar characteristics via different developmental processes that depend on different genes and their environments. Thus, several different alleles have been found to be associated with dark and light skin pigmentation in different populations. It is also clear that differences in skin color are not due to strong selection related to a few alleles.[2]

[1] Jablonski, N. G., and Chaplin, G. (2010). Human skin pigmentation as an adaptation to UV radiation. *Proceedings of the National Academy of Sciences, 107*(Supplement 2), 8962–8968.

[2] Jablonski, N. G. (2021). The evolution of human skin pigmentation involved the interactions of genetic, environmental, and cultural variables. *Pigment Cell & Melanoma Research, 34*(4), 707–729; Rocha, J. (2020). The evolutionary history of human skin pigmentation. *Journal of Molecular Evolution, 88*(1), 77–87.

62 ANCESTRY REIMAGINED

and Beyonce. Therefore, the category "Black" does not naturally exist in the world, but is rather socially constructed on the basis of our representations of it. Following philosopher Ron Mallon, we can thus state that "Black" (or any other race or social group for that matter) is socially constructed if and only if the existence of this category is caused by human mental states, decisions, and cultures of social practices. This means that the category "Black" depends on our mental states, and social practices, and would have not otherwise existed (I take "social" as a shorthand for "mental/social/cultural").[27]

One of the best illustrations of the social construction of race is what came to be known as the "one drop rule" in the US South. This is the idea that even a single drop of "Black blood" would make a person Black; this entails that anyone who has any known African ancestry is considered to be Black—independently of their skin color. According to the "one drop rule," any mating with White people would never liberate the descendants of Black people from their "blackness"; whereas White people would compromise the "whiteness" of their descendants if they mated with Black people. This "rule" is perhaps best illustrated by the notion, common in the United States, that whereas a White woman could give birth to a Black child, a Black woman could never give birth to a White child.[28] Consider an example: former US President Barack Obama is often described as the first African American president. But here is how Obama has described himself: "I am the son of a black man from Kenya and a white woman from Kansas."[29] Why do we think of Obama as a Black person with a White mother, rather than as a White person with a Black father? One reason is the assumption that if a person has some Black ancestry, it qualifies him as Black. In short, race is a social construct because there are different criteria for distinguishing among races, and we decide which ones to use (e.g., it can actually be African ancestry rather than skin color).

For another illustration of the social construction of race, consider the categories in the US Census. The US Census Bureau has the mission to provide data about the people and the economy of the country. They consider information about race as critical for making policy decisions, for promoting equal employment opportunities, and for assessing racial disparities in

---

[27] Mallon, R. (2016). *The construction of human kinds.* Oxford: Oxford University Press, pp. 1–6.

[28] This is why this rule was also described as the "hypo-descent rule" by anthropologists, which meant that a racially mixed person would be assigned the status of the subordinate group. See Davis, F. J. (2010). *Who is Black? One nation's definition.* University Park: The Pennsylvania State University Press, pp. 5–6, 11–14.

[29] https://www.huffpost.com/entry/obama-race-speech-read-thn92077

health and environmental issues. Their statement about race notes: "The racial categories included in the census questionnaire generally reflect a social definition of race recognized in this country and not an attempt to define race biologically, anthropologically, or genetically. In addition, it is recognized that the categories of the race item include racial and national origin or sociocultural groups."[30] For instance, people from China and India, who represent about half of the world population and who can be perceived to differ in their visible features, including skin color, are often lumped together in the racial category "Asian." But perhaps the best piece of evidence that race is a social construct is how the different races, and their descriptions, included in the US Census have changed across time due to the public attitudes and the politics of each period. For example, "colored," "Black," "Negro," and "African American" have been used at different times, with the term "Negro" eventually being dropped for the 2020 Census.[31] I must note that until 1950, it was census-takers who determined the race of the people they counted. Since 1960, however, people have been asked to choose their own race, and since 2000 it has become possible to identify with more than one racial category.

If races were biological, there would be no change in the US Census categories between 1790 and 2020. Of course, one might claim that some racial categories have remained stable across time, such as the category "White." This is not accurate though; for instance, people of Italian origin were not always considered as White. Most importantly, we should not forget that "White" is the category with which the people who created this classification scheme identified, and what this might show is their attempt to clearly delineate their own group from others. But to achieve this, they constructed for themselves both their ingroup and the outgroups. Of course, having said that racial groups are socially constructed does not mean that they do not really exist. They do exist and they have an enormous influence on people's lives; but they are a human invention, not the outcome of our biology. Races are real, but they are not biological.

Perhaps because race has been a sensitive social issue, and certainly because DNA analyses have been able to locate a person's ancestry not only with respect to a continent but also with respect to regions within continents,

---

[30] https://www.census.gov/topics/population/race/about.html
[31] For interactive graphs and a complete presentation by decade, see https://www.census.gov/newsroom/blogs/random-samplings/2021/08/improvements-to-2020-census-race-hispanic-origin-question-designs.html (accessed January 20, 2022).

64 ANCESTRY REIMAGINED

DNA ancestry companies currently provide customers with what has been called ethnicity estimates. Let us then see what ethnicity is.

Ethnicity

In Chapter 1, I mentioned three reasons for focusing on ethnicity in the present book: (1) it is in practice a major, although not universal, criterion for distinguishing humans into social groups; (2) it is what DNA ancestry test results claim to reveal, and then I defined the notion of genetic ethnicities; and (3) it could be considered as more fundamental than race and nationality, because people of the same race or nationality may have different ethnicities. Perhaps a fourth reason, given what we have discussed in the present chapter, is that it can be perceived as a less sensitive and loaded concept than race. Already in 1950, the United Nations Educational, Scientific, and Cultural Organization (UNESCO) proposed to replace "race" with references to ethnic groups:

> National, religious, geographic, linguistic and cultural groups do not necessarily coincide with racial groups: and the cultural traits of such groups have no demonstrated genetic connexion with racial traits. Because serious errors of this kind are habitually committed when the term "race" is used in popular parlance, it would be better when speaking of human races to drop the term "race" altogether and speak of ethnic groups.[32]

Both the noun "ethnicity" and the adjective "ethnic" derive from the ancient Greek word *ethnos* (ἔθνος). In antiquity, *ethnos* simply denoted a class of (generally animate) beings who shared a common identification.[33] In modern Greek, *ethnos* (plural *ethnē*) refers to a group of people who want to distinguish themselves as such, on the basis of a long-term cohabitation of the same geographical region, a common history and culture, and a real or assumed phyletic homogeneity;[34] in other words, it is used to refer to what in English we refer to as "ethnic nation." However, the way "ethnic group" and "ethnicity" are used today is close to the meaning of the modern Greek

[32] https://unesdoc.unesco.org/ark:/48223/pf0000128291 (accessed July 16, 2022).
[33] I am grateful to Jonathan Hall for his suggestion to make this clear.
[34] https://www.greek-language.gr/greekLang/moderngreek/tools/lexica/triantafyllides/search.html?lq=ἔθνος&dq=

word *ethnotita* (εθνότητα), which has the features of an ethnos but is not recognized as being autonomous and is rather subsumed to a larger entity such as a nation.[35]

Ethnic identity and ethnicity have been recently brought to the fore in the Western world. One important reason is that immigration and globalization have resulted in a variety of clashes among different ethnic groups in very different contexts. A common aspect of many related discussions is the emphasis on the differences between the opposing groups, emphasizing the contrast between "us" and "them." This contrast enhances two negative views: ethnocentrism, considering one's own ethnic group as being the center of the world, and ethnic prejudice, being prejudiced against other ethnic groups. For instance, many immigrants strive to preserve their cultural identity, transforming areas in their host countries so as to resemble those in their country of origin. This in turn makes the people in the host countries feel threatened that they might lose their own ethnic identity. As a result, these tensions may enhance both their ethnocentrism and their ethnic prejudice. Of course, situations like these are not new. Colonization has historically been a main cause of multiculturalism for many countries. For instance, there are French nationals today whose parents or grandparents were born in French colonies in North Africa. Various kinds of inequalities and discrimination among ethnic groups have existed for a long time; the conflict between Hutu and Tutsi in Rwanda in the 1990s is one of many sad stories of this kind.

In principle, our ethnic identity is not our personal choice, but it is somehow imposed on us at birth, depending on the ethnic identity of our parents. I was born in Greece, all my parents, grandparents, and great-grandparents were Greek, and I inherited the language, the traditions, and the customs that were transmitted—culturally, of course—from one generation to the next. Therefore, I consider myself, and I am considered by others, as having a Greek ethnic identity. But being ethnically Greek was not my choice—I have absolutely no problem at all with this, but the case remains that it is not a conscious decision I made. It is due to where I was born and how I was brought up. Of course, our identities can evolve and change. But they will always leave some indelible signs. For instance, even though I can read and speak English and French, the language with which I can best and

---

[35] https://www.greek-language.gr/greekLang/moderngreek/tools/lexica/triantafyllides/search.html?lq=εθνότητα&dq=

66 ANCESTRY REIMAGINED

most naturally express myself is Greek. This will never change.[36] Historian Shlomo Sand described this feeling in a vivid manner in his personal account of how he rejected his own Jewish identity: "I inhabit a deep contradiction. I feel like an exile in the face of the growing Jewish ethnicization that surrounds me, while at the same time the language in which I speak, write and dream is overwhelmingly Hebrew."[37]

What is important about all this is that because ethnic identity is (culturally) inherited and leaves indelible signs such as one's language, traditions, and often habits of mind, we might be prone to confuse it with a genetic characteristic. Ethnic groups are descent communities because of cultural inheritance, but this goes in parallel with genetic inheritance. Therefore, people might end up thinking that an ethnic essence might exist that is transmitted along the way. This view might be enhanced by the marketing of DNA ancestry testing companies. As already mentioned in the Preface, these companies claim that their DNA tests can reveal your true ethnic identity. This can be perceived as an essentialist attitude that prescribes that identity is due to an inner essence—DNA—and so all we need to do is "read" it and figure out who we really are. Recall how certain Karen was about being 100% East African and Jay about being 100% English. When they received their DNA results that indicated otherwise, they were shocked because they took the DNA test results to reveal the truth about them to the point that they questioned what they already knew from their families.

Thinking in an essentialist way about ethnic groups is not new, of course. During the 19th century, ethnic groups were strongly essentialized and considered as being monolithic, bounded, and immutable—in the same manner with races. This is why the notion of origin was of utmost importance, and it was crucial to establish it as far back in time as possible. As a result, ethnic groups were often considered as extended families, and ethnicity was considered as somehow inheritable and able to determine behavior and culture independently of the context. In this perspective, described as primordialism, ethnic identities have a somehow biological component and are therefore stronger than other social identities.[38] During the 20th century,

---

[36] Coulmas, F. (2019). *Identity: A very short introduction*. Oxford: Oxford University Press, pp. 74–77.

[37] Sand, S. (2014). *How I stopped being a Jew*. London: Verso, p. 101.

[38] But why might we think like this? An interesting idea is that we think of ethnic groups in the same way we think of species as natural living kinds. The reason for this is that ethnic groups exhibit two crucial similarities with species: (1) their members tend to have offspring with members of the same category and (2) their being members of a particular category is due to their descent from ancestors who were themselves members of that category. See Gil-White, F. J. (2001). Are

FROM RACE TO ETHNICITY IN ANCESTRY TESTING   67

however, ethnic identities were redefined as dynamic and mutable. This perspective is described as instrumentalism, and it questioned the essentialist assumptions of primordialism. A key issue here is that ethnic groups cannot be clearly delimited or defined. There, of course, exist groups of people with common aspirations and ascriptions. However, they are neither fixed nor clearly distinct from other such groups for many reasons, such as intermarriage between people of different groups and migration. Still, many people continued to accept primordialism, and this is actually the perspective that often characterizes our perception of the continuity of ethnic groups since antiquity (a point to which I return in Chapter 9). However, both primordialism and instrumentalism face problems. On the one hand, because primordialism essentializes ethnicity, it tends to consider that the existence of ethnic groups is indicated by the distribution of cultural traits. However, the same cultural traits can be shared by different ethnic groups that interact with one another. On the other hand, instrumentalism cannot account well for the strong sentiments that people often have for their ethnic group. Why would people sacrifice themselves for something that is socially constructed and not essentially real? Nevertheless, today most scholars accept a dynamic perspective of ethnicity that emphasizes the mutability of ethnic identities, which are in turn constructed as a response to certain needs.[39]

Sociologist Rogers Brubaker has made the interesting suggestion that perhaps groups do not even really exist at all. He described "groupism" as "the tendency to take discrete, bounded groups as basic constituents of social life, chief protagonists of social conflicts, and fundamental units of social analysis." This is also the tendency to reify groups such as Serbs and Croatians in former Yugoslavia, Jews and Palestinians in Israel, or Turks and Kurds in Turkey "as if they were internally homogeneous, externally bounded groups, even unitary collective actors with common purposes. I mean the tendency to represent the social and cultural world as a multichrome mosaic of monochrome ethnic, racial or cultural blocs."

ethnic groups biological "species" to the human brain? Essentialism in our cognition of some social categories. *Current Anthropology, 42*(4), 515–553.

[39] For an overview of the development of scholarly ideas on ethnicity, see Smith, A. D. (2010). *Nationalism.* Cambridge: Polity, pp. 53–60; see also Siapkas, J. (2014). Ancient ethnicity and modern identity. In J. McInerney (Ed.), *A companion to ethnicity in the ancient Mediterranean* (pp. 66–81). Chichester: John Wiley & Sons.

68 ANCESTRY REIMAGINED

Instead of adopting such an essentialist view, Brubaker suggested that we had better think of ethnicity (as well as race and nation) not as an entity, but rather as a category. In other words, our focus should not be on a "group" as a collection of individuals, but on "groupness" as a property that varies with context. This entails that groupness can be perceived as an event: people may experience extraordinary cohesion and collective solidarity in the name of their common ethnicity. But this does not have to be a permanent situation, and indeed it may not happen at all. What then matters is not the group as a distinct collectivity, but the conditions for group formation. Eventually, ethnicity is not something that exists in the world, but it is a way to perceive, interpret, and represent the world. This means, for instance, that the Greeks are not a permanent, bounded, distinct group that exists in the world, but rather a category to which people can be ascribed on the basis of political decisions, cultural expression, and everyday experience.[40]

What does this entail? That ethnicity is not fixed, but rather permeable. As already mentioned, ethnicity is acquired at birth. A newborn not only will be considered initially as having the ethnicity of one or both of its parents; it will also grow up in a particular social and cultural context that will characterize their way of life. But this particular ethnicity will have to be asserted by that person. For instance, a boy born in Greece by Nigerian parents (NBA star Giannis Antetokounmpo is one example) may assert the ethnic identity of his parents and consider himself as ethnically Nigerian; or he may end up considering himself as ethnically Greek because he was born, grew up, and was schooled in Greece, and so has assimilated all Greek customs and traditions, as well as the language. Or he may consider himself as bi-ethnic, asserting both ethnic identities. However, ethnicity is also a matter of ascription. As the son of Nigerian parents, he is visibly different in terms of skin color from the other children born to Greek parents, and his classmates may not consider him as being Greek for this reason; or because he speaks Greek as fluently as they do, and participates in the same social and cultural activities with them, they may consider him Greek, independently of his skin color. And there is also the Greek state that might confirm his Greek ethnic

---

[40] Brubaker, R. (2004). *Ethnicity without groups.* Cambridge, MA: Harvard University Press, pp. 8, 11–24; Brubaker provided several examples to make his case, one of which has been studied extensively in detail and presented in Brubaker, R., Feischmidt, M., Fox, J., and Grancea, L. (2008). *Nationalist politics and everyday ethnicity in a Transylvanian town.* Princeton, NJ: Princeton University Press.

FROM RACE TO ETHNICITY IN ANCESTRY TESTING   69

identity by assigning him the Greek nationality or deny it on the basis that he was not born to Greek parents.[41]

This raises the most interesting question: what are the criteria for ethnicity? Political scientist Donald Horowitz has suggested a very useful distinction between "the criteria of identity, on which judgments of collective likeness and unlikeness are based, and the operational indicia of identity, on which ready judgments of individual membership are made." This means that there is a *definitional set of attributes*, the criteria, on the basis of which it is decided whether a person has a particular ethnicity. These criteria are the result of conscious choices that attach significance to some attributes, descent being the most important one for ethnicity. There are also the indicia, which are an *operational set of attributes*, from which we can infer a person's ethnicity but which do not determine it. In other words, the indicia are evidence of an ethnicity, and they are developed only after the criteria have been decided. Indicia have a high probability to correctly assign a person to an ethnicity, but they can also fail to do so. Criteria and indicia can be confused when indicia are used for a long time and eventually start being considered as criteria.[42] To give a simple example, the fact that all my ancestors (as far as I know) were born and lived in Greece, and were themselves Greek, is the criterion for my own ethnicity also being Greek. The fact that I speak and read Greek is one of the indicia for my Greek ethnicity. Someone with these features is very likely to be Greek, but this is not the defining criterion as some non-Greeks also speak and read Greek.

This distinction between criteria and indicia for ethnicity brings us back to DNA ancestry testing.

DNA Markers Are Indicia, Not Criteria, for Ethnicity

I have already mentioned that DNA ancestry testing results refer to groups at the subcontinental level. Thus, as we saw in the Preface, Karen and Jay were not only told that they were African and European, respectively—this is something they knew anyway; they were also assigned to groups related to particular geographic regions with their continents of origin. In doing so,

[41] Coulmas, F. (2019). *Identity: A very short introduction*. Oxford: Oxford University Press, pp. 29–33.

[42] Horowitz, D. L. (1975). Ethnic identity. In N. Glazer and D. P. Moynihan (Eds.), *Ethnicity: Theory and experience* (pp. 111–140). Cambridge, MA: Harvard University Press, pp. 119–120.

70 ANCESTRY REIMAGINED

geography, culture, and DNA are merged and result in the categories that in Chapter 2 I described as genetic ethnicities. The implicit assumption is that there are particular categories, and identities stemming from them, which define groups in an essentialist manner: setting strict boundaries, distinguishing between different types of people, thus prescribing criteria for inclusion and exclusion in them. But this is utterly wrong and misleading. Let us see why.

As I explain throughout the present book, there are some markers on DNA that are found more often in some groups than others. Once you take a test and the company figures out which markers you have, you can be assigned to one or another group. But it is important to realize that these DNA markers are indicia, not criteria for ethnicity. They are an operational set of attributes, from which we can (probabilistically, as I explain later) infer a person's ethnicity but which do not determine it. Therefore, Karen and Jay are not what their DNA tests indicated because the DNA markers tested are not the criteria for their ethnicity. Karen and Jay could have the ethnicity that their tests indicated, under particular assumptions that I explain in Chapters 6–9. That Karen could be mostly Southeast African and Jay could be mostly Irish are just inferences based their DNA markers that are indicia, not criteria for ethnicity. Therefore, there was absolutely no reason for Karen and Jay to feel stressed or reconsider the family stories and the ethnic identity their parents and grandparents had told them they had.

This does not mean that the underlying science is wrong. This is not the case at all, and indeed as I explain in the last chapter, it can be very useful for finding relatives with whom people share a relatively recent common ancestry. Because social groups are socially constructed, there is no way that DNA markers can be a criterion about whether a person belongs to one or the other group. However, it is possible that DNA markers might indicate this (this is what indicia means). If we have identified some markers that are found more often in one group than in others, it is possible to (probabilistically, as I explain later on) infer a person's membership in a group insofar as this person is found to have the respective markers. In this sense, the DNA markers are indicia for a person's ethnicity. For instance, if we find that I carry DNA markers usually found in Greeks (an explanation of how such a reference group is defined will have to wait until Chapter 10), these markers can be perceived to indicate that I am Greek. But if we do not find me having these markers, it does not in any way mean that I am not Greek, because the criteria for my ethnic identity lie in culture, not my DNA.

Nor does this mean that if those markers are found in a person who has no known connection to Greece, this person must be Greek. This might be the case, but it also might not.

The main reason for this discrepancy is that not all of our ancestors are reflected in our DNA. DNA molecules are transmitted across generations from ancestors to descendants, and thus form the genetic tie among them. But not from all ancestors to a descendant. To understand why, it is necessary to distinguish between two main types of ancestry: genealogical ancestry and genetic ancestry.

4

Genealogical and Genetic Ancestry

Ancestors and Descendants

The commercials of ancestry companies emphasize the importance of family connections. For instance, in one commercial, actor Rob Lowe is looking at his family tree on Ancestry.com with his two sons. The commercial states: "The holidays are all about family, and this year many families are finding new ways to connect. Watch as @RobLowe, with a little help from his sons, explore[s] his passions for history, family, and storytelling to create a connection with a distant relative through Ancestry*." Lowe himself also notes the importance of this kind of activity: "This is the new campfire; this is the new place where we gather, and we share and we discover."[1]

The main message of this commercial is that finding out more about your family gives you a better sense of who you are, where you come from, and, therefore, where you belong. Families consist of people to whom we are related through close connections, which are generally described as kinship. Kinship conventionally refers to relationship between persons either through descent, which is described as consanguineal, or through marriage, which is described as affinal. Therefore, persons such as one's parent, sibling, cousin, grandparent, or grandchild are described as consanguineal relatives, whereas persons such as one's father-in-law, mother-in-law, sister-in-law, or brother-in-law are described as affinal relatives.[2] Kinship relations based on descent are sometimes considered the most important ones. For instance, the participants in a survey in the United States[3] expressed a greater sense of obligation in a time of need toward relatives than nonrelatives. More participants expressed an obligation to help out their parents (83%), a grown child (77%), their grandparents (67%), or their siblings (64%) than the parent of a spouse or partner (62%), a grown stepchild (60%), a stepparent (55%), a step or half

---

[1] https://twitter.com/Ancestry/status/1330873807154933763 (accessed February 28, 2022).
[2] Stone, L., and King, D. E. (2018). *Kinship and gender: An introduction.* New York: Routledge.
[3] It involved a nationally representative sample of 2,691 adults in the United States, and it took place in October 2010.

Ancestry Reimagined. Kostas Kampourakis, Oxford University Press. © Oxford University Press 2023.
DOI: 10.1093/oso/9780197656341.003.0004

GENEALOGICAL AND GENETIC ANCESTRY 73

sibling (43%), or a best friend (39%).[4] These results indicate that kinship being based on biology, reproduction, and heredity is privileged.

However, kinship relations between parents and children are not necessarily biological; in assisted reproduction clinics where donor sperm is used, and in adoption agencies, families have emerged whose members do not necessarily all have a genetic relation.[5] That reproduction requires the fusion of sperm and egg, and that DNA is the basis of genetic inheritance, are biological facts. Nevertheless, biological facts alone do not tell us anything about kinship; it is their study, how we interpret it, and what conclusions we draw from it that matter. Recent work has shown a tendency toward a problematization of kinship, families, and blood relations. A variety of novel and alternative modes of creating persons and creating relations among them have supplemented family relations.[6] In spite of this, many people still privilege kinship relations that are biological. These have often been described as "blood" relations, meaning that there is something inherent in us that is shared with our kin and that results in a strong connection among us. This is a rather essentialist view, as it assumes that there is something inside— our common biology—that connects biological relatives. In this view, you will always be connected to them—whether you like it or not—whereas friendships, marriages, and any other kind of personal relations may one day come to an end. This view is perhaps best illustrated in the various cases of people who found out after a DNA test that the man they regarded as their father was not in fact their biological father—recall the case of novelist Dani Shapiro in Chapter 1. If they finally find out who their biological father is and meet him, they may feel that they have at last managed, as Shapiro described her personal experience in her autobiographical book, "to move forward, to fill in the missing pieces."[7]

The most fundamental genetic relations are parenthood (being a parent of someone) and filiality (being the offspring of someone), because all other

[4] https://www.pewresearch.org/social-trends/2010/11/18/the-decline-of-marriage-and-rise-of-new-families/ (accessed June 3, 2021).

[5] Franklin, S., and McKinnon, S. (2001). Introduction: Relative values: Reconfiguring kinship studies. In S. Franklin and S. McKinnon (Eds.), *Relative values: Reconfiguring kinship studies* (pp. 1–26). Durham, NC: Duke University Press.

[6] Franklin, S. (2001). Biologization revisited: Kinship theory in the context of the new biologies. In S. Franklin and S. McKinnon (Eds.), *Relative values: Reconfiguring kinship studies* (pp. 302–325). Durham, NC: Duke University Press, ; Franklin, S. (2013). *Biological relatives: IVF, stem cells, and the future of kinship*. Durham, NC: Duke University Press; Featherstone, K., Atkinson, P., Bharadwaj, A., and Clarke, A. (2020). *Risky relations: Family, kinship and the new genetics*. New York. Routledge.

[7] Shapiro, D. (2019). *Inheritance: A memoir of genealogy, paternity, and love*. London: Daunt Books, p. 290.

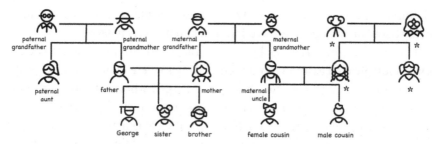

Figure 4.1 Family relations. George has genetic connections to all the people indicated in the figure, with the exception of those marked with an asterisk who are considered as related to him through marriage. (Free Icons from the Streamline Icons Pack)

relations are practically extensions of these. For instance, a grandparent is the parent of one's parent, and conversely a grandchild is the child of one's child. The unit that we can use to refer to genealogical time and to measure genealogical distances vertically between ancestors and descendants is the number of generations. The length of this unit in years can vary, depending on what one assumes to be the average number of years per human generation (often called "generation interval"; I call it "generation time" for simplicity): assuming that the average age for becoming a parent is 30 years old, then the generation time would be 30 years.[8] Parents and children belong to two different, successive generations, and so they are one generation apart. In this sense, a person's grandparents are found two generations back, that person's great-grandparents three generations back, and so on.[9]

Let us imagine that we are interested in the relations of a fictitious person called George, as shown in Figure 4.1. The figure presents a portion of George's wider family. It includes persons who have a genetic connection to George as well as others (the persons with an asterisk) who do not. A genetic connection in this case means that they have a very recent common ancestor (parent, grandparent, great-grandparent) with George. This is not the case for the persons with an asterisk who are George's (affinal) relatives through the marriage of his maternal uncle. Interestingly, these people are

---

[8] Fenner, J. N. (2005). Cross-cultural estimation of the human generation interval for use in genetics-based population divergence studies. *American Journal of Physical Anthropology*, 128, 415–423.

[9] For a detailed discussion of genealogy concepts, see Zerubavel, E. (2012). *Ancestors and relatives: Genealogy, identity, and community*. New York: Oxford University Press.

genetically related to George's first cousins, who are also genetically related to him. However, these two kinds of genetic relations are not linked: George is genetically related to his first cousins through his mother, his maternal grandparents, and his maternal uncle, whereas his first cousins are related to the persons with an asterisk through their own mother who has no recent genetic relatedness to George. The important point here is that whereas all persons presented in Figure 4.1 have a family relation to George, not all of them have a genetic relation to him through recent ancestry. Some are consanguineous relatives, and some are affinal relatives.

That "blood" relations are considered important is a main reason that people are interested in DNA ancestry testing. But why we would even think like this? As sociologist Eviatar Zeruvabel explained:

> Yet as implicit in our notion of "blood ties" (and of being "related by blood") and our vision of lineages as "bloodlines," we had already biologized and thereby essentialized ancestry and kinship long before we discovered genetics. Indeed, we have long viewed blood as what actually allows familial, ethno-national, or ethnoracial "essence" to be transmitted from ancestors to their descendants.[10]

In other words, we have long thought of blood relations in essentialist terms, before genetics. What genetics may have done is to materialize these essences in the form of DNA.

Starting from parents and offspring, one can theoretically trace all of one's ancestors for as long as relevant genealogical information is available. The result can be a network of ancestors and descendants connected to one another through successive generations. Unfortunately, except for some rare cases where detailed records have been kept for long periods of time—a notable case are the members of The Church of Jesus Christ of Latter-day Saints, known as the Mormons, who have since the 19th century strived to collect records of genealogical importance[11]—the available information is more limited, the further back one goes into the past. One can certainly find information from the family stories transmitted across generations, but usually

---

[10] Zerubavel, E. (2012). *Ancestors and relatives: Genealogy, identity, and community.* New York: Oxford University Press, p. 55.
[11] See Creet, J. (2020). *The genealogical sublime.* Amherst: University of Massachusetts Press.

one does not have first-hand testimonials beyond one's great-grandparents. But many ancestors have existed before them.

The number of our ancestors doubles in each generation as we go back in time, because every person in any generation has had two parents. My parents had two parents each, and so I had four grandparents; those in turn had two parents each, and I thus had eight great-grandparents. That each one of us had eight ancestors three generations back can also be expressed in mathematical terms, given that each ancestor had two parents: $2 \times 2 \times 2 = 2^3$, or 8 great-grandparents. If we went n generations back, each one of us would have had 2^n ancestors. Assuming a generation time of 30 years on average, 300 years or 10 generations back, there have existed (in theory) 2^{10}, or 1,024, ancestors for each one of us. But if we go 40 generations back, or 1,200 years ago, the number of ancestors for each one of us becomes more than a trillion. This is more than the total number of people who have ever lived, which is estimated to have been a bit more than 100 billion.[12] And how about the trillions of ancestors of other people? Is this possible?

The answer is that the calculation I have presented assumes that our ancestors who have lived in the past in every generation were different individuals. But this is not actually so. Rather, as we go back in time, some parts of our family tree overlap as some of our ancestors may have had common ancestors themselves. For instance, if a couple are sixth cousins— something likely if their ancestors originated from the same area, and something they might not even be aware of—this means that they had the same great-great-great-great-great-grandparents. Therefore, our family tree does not continuously expand as we go back in time. Most importantly, and for the same reason, our family tree would partially overlap with the family trees of other people—the further back we went in time, the more significant this overlap would be—simply because there would be many common ancestors. Our family tree extended back in time would look like an extremely complex web with many overlapping nodes (ancestors) and lines (relations) connecting them.[13] This phenomenon is described as family tree "collapse," because distinct family trees can no longer be sustained, and it has two important consequences.

The first consequence is that it makes no sense to talk about bloodlines, ancestral lines, lines of descent, and so on. There are no singles lines in the

---

[12] https://www.prb.org/howmanypeoplehaveeverlivedonearth/
[13] See Relethford, J. H., and Bolnick, D. A. (2018). *Reflections of our past: How human history is revealed in our genes.* New York: Routledge, p. 215.

past; it is family trees that exist, and there is no real reason that some of these lines should be privileged compared to others. However, some people like to stretch their genealogical research back in time, trying to find famous ancestors around whom they might construct their personal stories and identities. For instance, English actor Christopher Lee once told an audience in University College, Dublin, about his music album related to Charlemagne, and added: "He was, in fact, my ancestor and we can prove it."[14] Charlemagne, born around 742 CE, was a medieval emperor who ruled much of Western Europe from 768 CE until his death in 814 CE.[15] Lee was born in 1922 and died in 2015, so he and Charlemagne lived around 1,200 years, or 40 generations, apart. Therefore, if Charlemagne was actually an ancestor of Lee, he was one of Lee's 2^{40} theoretical ancestors, which is an inconceivably huge number. Was Charlemagne the ancestor of any other people living today? The answer to this question has to wait until the end of this chapter, because we first need to understand what exactly connects ancestors and descendants.

The second important consequence of the phenomenon of family tree collapse is that even though the persons with an asterisk in Figure 4.1 are not closely related to George in the way the family is depicted in that figure, this does not mean that they are not at all genealogically related to him. It could be possible that these persons had a not-so-distant common ancestor with George who would be included in a larger family tree that would extend more than two generations back. In fact, the further back in time the tree went, the larger it would be and the broader its coverage would be. Therefore, at some point we might find the common ancestors of the persons with an asterisk in Figure 4.1 and George. This entails that there is no natural way in which we can set boundaries in a family and clearly distinguish between relatives and nonrelatives. Whereas it is certainly the case that some relatives are genealogically closer to George than others because their common ancestor with George is more recent, those other people whose common ancestor with George is more distant are still relatives—they are just more distant ones. Therefore, families are inherently boundless, and any boundary we might set would be a matter of social convention alone and thus abstract.[16]

---

[14] https://www.youtube.com/watch?v=9Tdl021ArsM (accessed January 27, 2022).
[15] https://www.history.com/topics/middle-ages/charlemagne (accessed January 27, 2022).
[16] Zerubavel, E. (2012). *Ancestors and relatives: Genealogy, identity, and community.* New York: Oxford University Press, pp. 71–72.

78 ANCESTRY REIMAGINED

I should note at this point that George's ancestors have a prominent place among his others relatives: his brother, his sister, his uncles, his aunts, and his first cousins. In all these cases, George and his relatives are descendants of a couple of common ancestors, and so it is through that couple that they are connected to one another. Therefore, George, his sister, and his brother are connected through their parents, who are their common ancestors. His maternal grandparents are the common ancestors that George shares with his maternal uncle and his first cousins; and his paternal grandparents are the common ancestors George shares with his paternal aunt. We can also use the number of generations to measure genealogical distances horizontally in terms of the number of generations separating co-descendants from their most recent common ancestors. For instance, George's first cousins are connected to him through their common grandparents, and therefore their connection goes two generations back.

What DNA ancestry tests can do is to help us find people with whom we are related through common ancestors. But are all our ancestors represented in our DNA?

"Ghost" Ancestors

The concept that describes the relations between ancestors and descendants is ancestry. But there are two different conceptions of ancestry that can be easily conflated, even though they differ from each other: genealogical ancestry and genetic ancestry.[17] Genealogical ancestry refers to the identifiable ancestors in a person's family tree (as in Figure 4.1). Usually, these are one's parents, grandparents, and great-grandparents, but it is possible that someone finds information about ancestors further back in time through written records or other sources. Genetic ancestry does not refer to one's family tree, but to the paths through which one's DNA was inherited. In fact, we inherit about half of our DNA from each of our parents. As a result, only a portion of our DNA comes from each of our ancestors: 50% from each of our parents, 25% on average from each of our grandparents, and 12.5% on average from each of our great-grandparents. But which 50% of their parents' DNA two siblings will inherit is a matter of chance. It depends on which chromosomes (Box 4.1) will be included, during meiosis (Box 4.2), in the

---

[17] See Mathieson, I., and Scally, A. (2020). What is ancestry? *PLoS Genetics, 16*(3), e1008624.

Box 4.1 What Are Chromosomes?

DNA does not exist on its own in the nucleus of each cell. Rather, it is part of chromatin, a molecule that results from the chemical combination of DNA and particular proteins called histones, thus forming a chromatin filament. Before cell division, DNA replication takes place, and each DNA molecule/chromatin filament doubles. During cell division, chromatin is condensed and forms chromosomes, with a characteristic X-like shape consisting of two identical chromatids. Both males and females have 46 chromosomes in each of their body cells. These 46 chromosomes are not entirely different from one another; rather, there exist chromosomes that contain very similar DNA sequences and thus form pairs of homologous chromosomes. Each chromosome of each pair is inherited from one of each person's parents. Among these chromosomes, 44 chromosomes, or 22 pairs, are similar in both sexes and are called autosomes. The 23rd pair is the pair of the sex chromosomes, which is XX in females and XY in males. Organisms and cells that contain such pairs of chromosomes are described as diploid (meaning double kind). The reproductive cells (sperm and ovum) that each contained these 23 chromosomes (one from each pair) are described as haploid (meaning single kind).

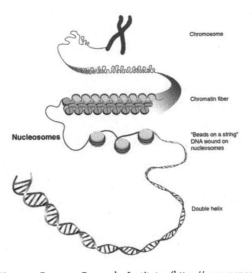

Source: National Human Genome Research Institute (https://www.genome.gov/genetics-glossary/Nucleosome).

Box 4.2 What Is Crossing Over?

In most cells of our body, cell division takes place in a way that each descendant cell ends up having the same chromosome number with each other and with the initial cell. Before cell division, DNA replication takes place. Thus, each DNA molecule, and each chromatin filament to which it belongs, comes to exist in pairs. During the division that is called mitosis, all double-filament chromosomes are aligned in the middle of the cell and one of the two chromatids of which they consist (each of which in turn consists of one DNA molecule and proteins) ends up in each of the two descendant cells. A cell with 46 chromosomes is thus divided into two cells that also have the same 46 chromosomes each (the number is maintained thanks to the replication of DNA).

However, in the ovaries and testes a special kind of cell division, called meiosis, takes place that results in the production of reproductive cells, sperm and ova, each of which has 23 chromosomes. After the replication of DNA in meiosis, the homologous chromosomes are put one opposite to the other. A first division takes place that brings one chromosome of each pair in each of the two resulting cells. Then a subsequent division occurs, during which the two chromatids of each chromosome are separated. Thus, four new cells emerge that have 23 chromosomes each. Now, the complication in meiosis is that autosomes are not static, but can actually exchange segments between them. What happens is that there is an exchange of parts between two homologous chromosomes (called crossing over), which results in new combinations of DNA segments in offspring that did not exist in their parents (called chromosome recombination). This has a very important consequence: we do not receive intact chromosomes from each of our parents, but rather chromosomes that are actually a combination of different chromosome segments from our parents. Let us consider a simple example. Imagine that the cells of my father contain a pair of chromosomes, each derived from one of my grandparents. Let's imagine that my father's cell contains a pair of chromosomes as in the embedded figure. One chromosome comes from my paternal grandfather, with the regions GFa and GFb, whereas the other chromosome comes from my paternal grandmother, with the regions GMa and GMb. Ga and Gb are described as different loci (singular locus) on the same chromosome. Several different alleles can exist on the same locus, in this case GFa and GMa for Ga, and GFb and GMb for Gb.

Each of the alleles that an individual has comes from each of one's parents, as is the case here. As you can see in the figure, if crossing over occurs, it is possible for spermatozoa to emerge that contain chromosomes that consist of segments of both of my grandfather's and my grandmother's chromosomes (c2, c3)! Because of the phenomenon of crossing over, our autosomes are not transmitted intact across generations, but in various combinations of chromosome segments.

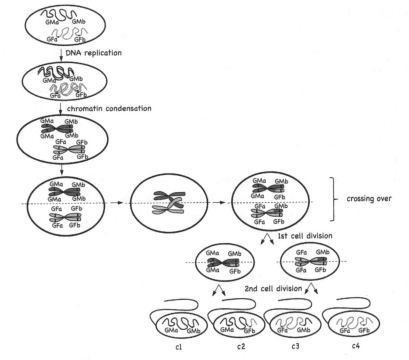

I must note here that four cells are actually produced in the meiosis, leading to the production of spermatozoa. In the ovaries it is actually one ovum that is produced, along with two polar bodies that do not divide further. This is why women have one ovum only available in each menstrual cycle.

reproductive cells, or gametes (sperm and ovum), from which an embryo will emerge. As a result, whereas George and his siblings have exactly the same genealogical ancestry and the same family tree (Figure 4.1), their genetic ancestry is different because they have each probably received different portions of DNA from their ancestors. This is why genetic ancestry may provide information about one's genealogical ancestry, but not the other way around.

Figure 4.2 shows the inheritance of chromosome 1 segments across four generations, from George's great-grandparents to him (we can ignore recombination for simplicity; we can imagine that these chromosome segments are transmitted intact across these generations). If George's great-great-grandparents had sixteen distinct chromosome 1 segments, labeled 1a, 1b, and so on, George has inherited only two of those because each person has inherited one from each of their parents. Thus, the probability for George receiving any two of these sixteen segments is 2/16 or 1/8. Therefore, even though three generations ago George has eight genealogical ancestors, only two of them will also be his genetic ancestors for the particular chromosome 1 segments: his father's maternal grandmother, from whom he inherited segment 1h, and his mother's maternal grandfather, from whom he inherited segment 1m. All people in Figure 4.2 are George's genealogical ancestors, but only the highlighted ones in gray are also his genetics ancestors for the particular chromosome 1 segments. Similar would be the case for any other segment from each of the twenty-two pairs of autosomes that George has (with the exception of his Y chromosome, which we consider in the next section).

Therefore, DNA inheritance is somehow "diluted" across generations. This also becomes evident if we compare ancestors and descendants in terms of the amount of shared DNA. George's children will have about 50% of their

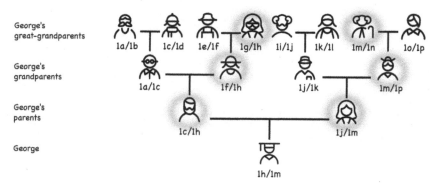

Figure 4.2 The inheritance of chromosome 1 segments across four generations, from George's great-grandparents toward him. From the 16 distinct chromosomes 1 segments that his great-grandparents had, George has only inherited two. Whereas all people in this family tree are George's genealogical ancestors, only those highlighted in gray are also his genetic ancestors for these particular chromosome 1 segments. (Free Icons from the Streamline Icons Pack)

GENEALOGICAL AND GENETIC ANCESTRY   83

DNA common with him; his grandchildren will also have 50% (0.5) of their DNA common with his children, and therefore 25% of their DNA (on average) common with George ($0.5 \times 0.5 = 0.25$ or 25%). Similarly, George's great-grandchildren will have 12.5% of their DNA (on average) in common with him ($0.5 \times 0.5 \times 0.5 = 0.125$ or 12.5%). And so on. In other words, the members of each generation of George's descendants will share 50% of their DNA with the previous generation; therefore, for any number of generations k, the members of the kth generation will share 0.5^k of their DNA with him. For instance, after seven generations, George's descendants in that generation will share 0.5^7, or 0.78125%, that is, less than 1% (on average) of their DNA with him.

The number of crossing-over events occurring in each meiosis has been estimated to be about 43 for mothers and about 28 for fathers. This is a total of 71 events per generation.[18] Therefore, since k generations back, there would have emerged $44 + 71 \times k$ chromosome segments from all of our ancestors in that generation. Consider George's two parents, who carry chromosomes from their own fathers and mothers, some of which have undergone recombination. In this case, there will exist 115 chromosome segments ($44 + 71 \times 1 = 115$). If we go back another generation, there has been another set of 71 recombination events, and so 186 chromosome segments will have existed since George's four grandparents. Up to nine generations back the number of segments is larger than the number of ancestors, and so it is likely that George has inherited a DNA segment from all of his genealogical ancestors at that generation. However, if we go 10 generations back, this is no longer the case. Whereas the number of distinct chromosome segments that would have emerged since then would be $44 + (71 \times 10) = 754$, the number of George's ancestors would be $2^{10} = 1,024$. As the number of chromosome segments is lower than the number of ancestors, George will not have inherited any chromosome segments from several of his 1,024 ancestors from 10 generations ago. Eventually, the more generations we go back in time, the higher is the number of ancestors who have contributed no DNA at all to George.[19]

In short, as we go back in time, the number of our ancestors increases, because it doubles in every generation, and the amount of DNA that we

---

[18] Chowdhury, R., Bois, P. R. J., Feingold, E., Sherman, S. L., and Cheung, V. G. (2009). Genetic analysis of variation in human meiotic recombination. *PLoS Genetics*, 5(9), e1000648.

[19] See Donnelly, K. P. (1983). The probability that related individuals share some section of genome identical by descent. *Theoretical Population Biology*, 23, 34–63; Edge, M. D., and Coop, G. (2020). Donnelly (1983) and the limits of genetic genealogy. *Theoretical Population Biology*, 133, 23–24; see also https://gcbias.org for answers to various questions related to ancestry.

84 ANCESTRY REIMAGINED

have inherited from them decreases, because it halves in every generation.[20] This results in the important difference between genealogical ancestry and genetic ancestry. If we go back 600 years, or 20 generations, repeating the calculations we did earlier, we find that we would have 1,048,576 ancestors since whom 1,464 chromosome segments would have emerged because of crossing over. The probability to inherit a chromosome segment from one of them would be, if you divide the two numbers, approximately 0.1%, or 1 in 1000. This means that there exist individuals who lived in the past who are at the same time genealogical ancestors of individuals living today, and genetic ancestors to none of them. These have been described as "ghost" ancestors.[21] After 20 generations, most genealogical relations are practically undocumented in the autosomal DNA data. This is something very important to keep in mind for those who hope to find their "deep ancestral roots" via DNA testing.

Let us then see about which of our ancestors' DNA can indeed provide information.

From Similarity to Ancestry

Despite the fact that different DNA segments may have different paths of inheritance, some of these are derived from the same common ancestor in people who are relatives. These segments are identical, or almost identical, with respect to a particular set of single nucleotide polymorphisms, because they have not been broken down by crossing over. For this reason, they are described as being identical by descent (IBD) segments.[22] The lengths of these IBD segments are measured in centiMorgans (cM), with 1 cM roughly corresponding to a DNA sequence of 1 million base-pairs (1 Mb; see also Box 1.1).[23] Closely related individuals will share larger IBD segments compared

[20] The situation becomes even more complicated if we take into account that closely related individuals, such as cousins, could have offspring. For instance, if two first cousins mate, whereas each one of them had four grandparents, because two of those were the same people, the total number of their ancestors two generations back is not eight but six.

[21] Gravel, S., and Steel, M. (2015). The existence and abundance of ghost ancestors in biparental populations. *Theoretical Population Biology, 101,* 47–53.

[22] Thompson, E. A. (2013). Identity by descent: Variation in meiosis, across genomes, and in populations. *Genetics, 194*(2), 301–326; Browning, S. R., and Browning, B. L. (2012). Identity by descent between distant relatives: Detection and applications. *Annual Review of Genetics, 46,* 617–633.

[23] One Morgan is defined as the chromosome distance over which one recombination event occurs. Thus, 1 cM corresponds to 1% probability that a marker on a chromosome will become separated from another marker on the same chromosome due to crossing over in a single generation; see https://www.genome.gov/genetics-glossary/Centimorgan (accessed January 27, 2022).

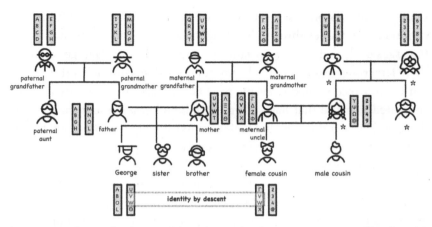

Figure 4.3 Identity by descent. The relationship between George and his female cousin can be established because they share an identical DNA segment that they have both inherited from one of their most recent common ancestors, their grandfather, through their mother and father, respectively, who are siblings. (Free Icons from the Streamline Icons Pack)

to more distantly related individuals, because there have been fewer recombination events since their common ancestor. The further back in time the common ancestor is found, the smaller the average length of IBD segments will be because more recombination events will have occurred since then. Very long IBD segments (larger than 80 cM) are very likely to come from an ancestor less than five generations back, whereas the majority of short IBD segments (up to 10 cM) likely come from an ancestor more than 20 generations back.[24]

Let us consider an example. As I have mentioned, even though George has inherited all of his DNA from his grandparents, via his parents, he does not have his parents' or grandparents' whole chromosomes. Rather, he has inherited particular segments from each one of them. As a result, he has particular segments in common with his siblings and his first cousins. Let us consider how a particular pair of chromosomes is transmitted across three generations. In order to refrain from making the illustration too complicated, I assume that only one recombination event occurs in one of the parents of each person. As shown in Figure 4.3, George and his first cousin

---

[24] Speed, D., and Balding, D. J. (2015). Relatedness in the post-genomic era: is it still useful? *Nature Reviews Genetics, 16*(1), 33–44.

86 ANCESTRY REIMAGINED

share the chromosomal segment VW, which they have inherited from their common grandfather through his daughter (George's mother) and his son (George's uncle and his cousin's father). At the same time, George will likely share larger segments on both chromosomes with his brother and sister because fewer recombination events have taken place (in their parents only) compared to those that have taken place between them and their cousin (both in their grandparents and their parents). Finally, it is important to note that because crossing over occurs repeatedly across generations, DNA from different common ancestors can end up on the same chromosome of an individual (for instance, one of George's chromosomes has the segment AB that existed in his paternal grandfather's chromosome, and the segment OL that existed as two different segments [O and L] in his paternal grandmother's chromosomes). It is in this way that DNA ancestry tests can find matches and potential relatives in their databases (see Chapter 10).[25]

There are two important exceptions to all of this: YDNA and mtDNA, which are inherited in a rather biased manner. YDNA is always transmitted from fathers to sons, because females do not have Y chromosomes (Box 4.3), whereas mtDNA seems to be always transmitted from a mother to her children (Box 4.4). As Figure 4.4 shows, George can thus trace his YDNA ancestry to a specific male ancestor across his father's lineage (the father of the father of the father, etc.) and his mtDNA ancestry to a specific female ancestor across his mother's lineage (the mother of the mother of the mother, etc.). Because mtDNA and YDNA, in contrast to autosomal DNA, do not undergo crossing over, they are transmitted intact across generations. Thus, George should have the same YDNA with his father's paternal grandfather (Y1 in Figure 4.4)—with whom he otherwise has only 12.5% (on average) of his autosomal DNA in common. Similar is the case for mtDNA. George has inherited his mtDNA from his mother's maternal grandmother (mtDNA4 in Figure 4.4), with whom he also has only 12.5% (on average) of his autosomal DNA in common.

So, given all this, how far back in the past can we identify our genetic ancestry? Population geneticists Peter Ralph and Graham Coop performed an analysis of genomic data for 2,257 Europeans. They found that the vast

---

[25] I should add here that having DNA "in common" only refers to the origin of the particular segment, not its sequence. It is possible that two people have the same DNA sequences, even though these were inherited from different ancestors. The DNA sequences compared for ancestry purposes are very variable, so that it is rather unlikely that people have the same DNA sequences due to chance and not due to common descent.

Box 4.3 What Is YDNA and How Is It Inherited?

Whereas a woman has two XX chromosomes that are homologous to each other, a man has an X and Y chromosome that are partially homologous to each other (the X chromosome is larger than Y). This happens because there are some regions on the Y chromosome that are similar to respective regions on the X chromosome. These are called pseudo-autosomal regions (PAR), because they are homologous in the same way that the autosomes are, and they are found at the two edges of the X and Y chromosomes. As a result, the middle region of the Y chromosome does not have a homologous one on the X chromosome.[1] This region is often called the male-specific region of the Y chromosome (MSY), or the nonrecombining region of the Y (NRY). For simplicity, hereafter I describe this region as YDNA, even though the DNA in the Y chromosome also includes the regions PAR1 and PAR2. YDNA is important because it contains DNA sequences that are implicated in the development of the male reproductive organs in the human embryo. The default developmental outcome for humans is to become females, and it is due to genes contained in YDNA (among others) that a switch to male development is possible. So, unless there are mutations, any person carrying YDNA is expected to be male. This is why we say that YDNA is transmitted from fathers to sons, resulting in what is described as patrilineal genetic inheritance. YDNA is haploid (see Box 4.1), does not undergo recombination, and so any differences among different YDNA molecules should be due to mutations only. It is the only part of DNA found within chromosomes that has this property, and it is about 56 Mb long. YDNA has high variation in its sequence, and it mutates faster than autosomal DNA.[2]

There are some exceptions to this binary view of sex, such as intersex and transexual people, who may have a karyotype that does not correspond to their gender. Some have raised discussions about the number of sexes that exist, but this topic falls outside the scope of the present book. For simplicity, given that the majority of male people are XY and the majority of female people are XX, in the present book they are considered as such.

[1] Flaquer, A., Rappold, G. A., Wienker, T. F., and Fischer, C. (2008). The human pseudoautosomal regions: A review for genetic epidemiologists. *European Journal of Human Genetics, 16*(7), 771–779.

[2] The mutation rate of YDNA seems to be somewhere between 89×10^{-9} and 76×10^{-9} mutations per base per year. Jobling, M. A., and Tyler-Smith, C. (2017). Human Y-chromosome variation in the genome-sequencing era. *Nature Reviews Genetics, 18*(8), 485.

Box 4.4 What Is mtDNA and How Is It Inherited?

Whereas most of our DNA is contained in the nucleus of each of our cells where our chromosomes are found, there is a very small portion of DNA outside that. This is found inside the mitochondria, which are small cylindrical organelles in the cytoplasm, the internal part of the cell outside the nucleus. Mitochondria are where chemical reactions occur that transform the energy contained in food to a form that cells can use, and they contain DNA, because they are the evolutionary descendants of bacteria. Mitochondrial DNA (mtDNA) contains two regions that are very variable and that are often used in ancestry identification, called hypervariable regions (HVRs: these are the HVR1 and HVR2). Each mitochondrion contains several copies of each mtDNA molecule, and each cell contains numerous mitochondria; therefore, any given cell contains numerous copies of its mtDNA, which is estimated to be about 0.3% of the total DNA of the cell.[1] The inheritance of mtDNA is described as matrilineal because we all inherit our mtDNA from our mothers. Whereas the sperm cells of our father also contain mitochondria, these do not enter the ovum during fertilization—only their DNA does. As a result, the fertilized ovum contains the DNA of both of our parents, but only the mitochondria that were already there and so only our mother's mtDNA. We thus see that YDNA and mtDNA are somehow sex-specific. It should be noted though that mtDNA is not as sex-specific as YDNA. YDNA is found only in males and is transmitted from fathers to sons. In contrast, mtDNA is found in both males and females, and it is transmitted from a mother to all her offspring, not just their daughters—sons also inherit their mother's mtDNA. However, we call it matrilineal because it will only be the daughters who will transmit it to their own offspring, even though their brothers have it, too.[2]

[1] The number of mtDNA copies varies from about one hundred in sperm to hundreds of thousands in an unfertilized oocyte. If we assume an average of 1,000 mtDNA molecules per human cell, given that each mtDNA molecule has a length of 16.569 Kb, we arrive at a total of about 16.569 Mb, which is about 0.3% of the diploid genome of 6.2 billion base-pairs.

[2] See Nash, C. (2015). *Genetic geographies: The trouble with ancestry*. Minneapolis: University of Minnesota Press, p. 145.

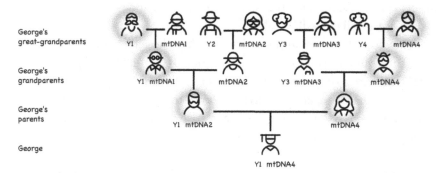

Figure 4.4 YDNA and mtDNA inheritance in George and his ancestors. (Free Icons from the Streamline Icons Pack)

majority of pairs compared (94%) shared only a single IBD segment. One general conclusion from this analysis was that the farther away the geographical locations were where two individuals lived, the lower the expected number of common ancestors—which makes sense as individuals living nearby are more likely to mate with one another than with individuals living far away. Another general conclusion was that pairs of individuals across Europe may share a common genetic ancestor within the last 1,000 years—1/32 on average for individuals living at least 2,000 km apart. However, as we go back in time, the probability increases. Thus, two individuals living at least 2,000 km apart are expected to share one common ancestor who lived between 1,000 and 2,000 years ago, and more than 10 common ancestors who lived between 2,000 and 3,000 years ago. This practically means that every person who has lived 1,000 years ago and has had descendants would be an ancestor of every European living today. In this sense, Charlemagne would not only be an ancestor of Christopher Lee but also my own ancestor as well as of everyone else born in Europe.[26] The further back we go in time, the more interconnected the family trees of each one of us become. It has been estimated that the common ancestor of all living humans, must have lived just a few thousand years ago.[27] This does not mean that there were no other people before that person, but that after a relatively recent point in time in

---

[26] Ralph, P., and Coop, G. (2013). The geography of recent genetic ancestry across Europe. *PLoS Biology, 11*(5), e1001555.
[27] Rohde, D. L., Olson, S., and Chang, J. T. (2004). Modelling the recent common ancestry of all living humans. *Nature, 431*(7008), 562–566.

90 ANCESTRY REIMAGINED

human evolution the ancestries of all humans living today converge. In other words, we are all related!

What does all this entail? Even if we privilege the genetic ties between ourselves and our ancestors, only a small portion of their DNA is represented in our DNA, and so only some of our genealogical ancestors are also our genetic ancestors. Therefore, at the individual level, there is no way that DNA can provide us with a complete view of our past. On the one hand, autosomal DNA can provide information for the totality of our genetic ancestry; but because of DNA recombination, it can only be about our "recent" (genetic) ancestry, that is, about common ancestors in the recent past. In contrast, YDNA and mtDNA can provide information about common ancestors hundreds or thousands of years ago, that is, our "deep" (genetic) ancestry; but this is information about a very small portion of our total genetic ancestry (around 1% combined). These are caveats we should keep in mind, along with the fact that as we go back in time there are more and more ancestors from whom we have inherited no DNA at all. What they entail is that autosomal DNA ancestry testing can provide some information for *some*, not all, of our *recent* genetic ancestors; whereas mtDNA and YDNA can provide information about more remote ancestors who correspond, however, to a minute portion of our genetic ancestry.

In order to better understand how autosomal DNA, mtDNA, and YDNA can provide information about ancestry, let us now consider some famous historical cases for which the comparison of DNA from ancestors, descendants, and relatives has provided some interesting insights.

5

Using DNA Ancestry Evidence
to Retrace History

Clarifying History Through DNA Ancestry

Nowadays it is possible for many people to find information about their recent ancestors. There are several limitations in this endeavor, as explained in Chapter 4, the most important being that not all of a person's genealogical ancestors are also their genetic ancestors. Yet, despite these limitations, DNA can provide very important and useful information in some cases, especially for people who may not have a deep and rich knowledge of their ancestors. This is especially important in cases of adoption, incorrect information or lack thereof, displacement due to politics or war, or migration. As a result, for many people DNA ancestry testing becomes the only means to seek answers to long-held questions about themselves. There are many such stories, so let us briefly consider one in order to have a concrete example of what kinds of questions people try to ask via DNA ancestry testing.

Daisy Fuentes is the host of the documentary series *A New Leaf*, airing on NBC in collaboration with Ancestry.[1] The aim of the show is to "follow people on their journey of self-discovery as they uncover their legacy. In each episode, these adventurers will begin to add new leaves on their family tree, opening up their life in a way they did not foresee."[2] The aim of the show is to add new, previously unknown leaves to one's family tree. One of the participants in his show was Heidi. Because her mother had been adopted, Heidi did not have much information about her family history on her mother's side. As she explained to the show host, "I've always enjoyed family history and [the] genealogy aspect and knowing why we're here, you know, I mean what made our lives possible." Heidi noted that she tried to find basic information about her mother's ancestry. She therefore petitioned

---

[1] https://www.ancestry.com/c/tv-shows/a-new-leaf (accessed July 3, 2022).
[2] https://www.youtube.com/watch?v=fYImnOiz2_8 (accessed July 3, 2022).

Ancestry Reimagined. Kostas Kampourakis, Oxford University Press. © Oxford University Press 2023.
DOI: 10.1093/oso/9780197656341.003.0005

92 ANCESTRY REIMAGINED

the court, but the response she received was that the records were sealed. Thus, she could not find anything. But she had questions about her biological grandfather, as well as about her family as a whole. There was also a story in the family that Heidi wanted to figure out if it was true: that her great-grandfather had been sold to a family in Germany to work on their farm, because his own family was poor.

Genealogical research managed to establish who exactly Heidi's maternal grandparents were, even revealing photos of them that Heidi had never seen. Her DNA ancestry results were 58% Germanic Europe and 22% Sweden, with the remaining 20% attributed to four other regions in Europe. Heidi was surprised by her Swedish ancestry. She was told by a genealogist that because in the late 1880s many Germans left to go to America, farm laborers from Sweden were recruited to work in the area that is today Germany. This could explain Heidi's Swedish ancestry. About the family story, Heidi said, "I was always told that my grandfather had been sold to a family in Germany, but it's looking more like that might not be the real story, that might have been just a family story and now that I've been able to discover the real story it's a feeling that I can't really describe." This is one of the numerous instances of excitement by people who come to learn more about their ancestry and their past. The combination of DNA data and genealogical information managed to provide answers to Heidi's questions.[3]

But, notwithstanding the excitement, Heidi's case points to two very important points. The first one is that whereas DNA data can be used to identify individuals with a high level of accuracy and some questions can be answered with a high degree of certainty, this does not mean that DNA data can provide answers to all questions a person may have. The second point is that DNA alone is not sufficient for providing answers. DNA can be used as data only when we can compare different DNA sequences, but we need to know which sequences we should compare. All interpretations of DNA data rely on evidence from additional historical and genealogical sources, which are extremely important. To illustrate these two points, in this chapter I describe three of some very early investigations in which DNA ancestry helped provide insights. These cases are presented not in a historical order but in order of increasing complexity, in order to make it easy for you to follow the science underlying them (Boxes 1.1, 1.2, 2.1, 4.1, and 4.2 may be useful to review while reading). The respective investigations aimed at clarifying

---

[3] https://www.youtube.com/watch?v=UtL9bJiLQJI (accessed July 3, 2022).

USING DNA ANCESTRY EVIDENCE TO RETRACE HISTORY   93

whether Martin Bormann, known as Hitler's executioner, escaped capture in Berlin and fled to South America; whether President Thomas Jefferson was the father of the children of Sally Hemings, an enslaved woman; and whether the remains found in a grave in the Urals area were those of Tsar Alexander II and his family who were allegedly executed in the aftermath of the Bolshevik Revolution.

Hitler's Executioner

Martin Ludwig Bormann was a powerful member of the Nazi Party and Hitler's personal secretary. Born in 1900, he joined the Nazi Party in 1927, and by 1933 he had managed to become Chief of Cabinet in the Office of the Deputy Führer, Rudolf Hess. He was always found around Hitler and in 1937–1938 he was behind the building of Hitler's "Eagle's Nest" mountain teahouse in the Obersalzberg area, which was made to impress foreign visitors, as it can be seen from almost anywhere in the surrounding area.[4] After 1941, when Hess flew to Scotland intending to convince the British to surrender, eventually being arrested and kept in prison until the end of the war, Bormann quickly rose in the Nazi Party as he had gradually managed to form a solid power base. By the end of 1942, he was Hitler's closest collaborator. As Hitler was preoccupied with military matters, he had to rely more and more on Bormann for domestic affairs, and eventually appointed him as his personal secretary in 1943. Bormann's views on race, the Jews, and forced labor made him Hitler's executioner, who made decisions about the fate of millions of people. Following Hitler's suicide in 1945, Bormann officially became head of the Nazi Party. He tried to flee Berlin as the Russian army was approaching, and he was never found alive. In 1946, he was sentenced to death by the International Military Tribunal in Nuremberg—in absence.[5] After the end of World War II, there were rumors that having anticipated the defeat of the Nazis, Bormann had planned, and eventually managed, to flee Europe for South America, where he coordinated an organization that relied on the wealth that the Nazis had managed to accumulate during the war.[6]

---

[4] Kaplan, B. A. (2011). *Landscapes of Holocaust postmemory.* New York: Routledge, pp. 18, 24.

[5] Koop, V. (2020) *Martin Bormann: Hitler's executioner.* Barnsley: Frontline Books.

[6] Manning, P. (1981). *Martin Bormann: Nazi in exile.* Secaucus, NJ: Lyle Stuart.

94 ANCESTRY REIMAGINED

What happened to Bormann was unknown until 1972 when two skulls were unearthed during construction works close to the Lehrter station in Berlin. That was the area where Bormann had last been seen alive in 1945, as witnessed by two of his companions, Artur Axmann and Erich Kempka. Therefore, it was possible that one of the two skulls was that of Bormann's. Whereas various methods were used for identification, eventually it was only the study of teeth that provided crucial evidence. This was possible because a Berlin dentist, Dr. Hugo Johannes Blaschke, had provided US army officers in 1945 with descriptions of the dental condition of Adolf Hitler, Eva Braun, and Martin Bormann. Blaschke had treated Bormann regularly since 1937 and so was able to give detailed diagrams of characteristic crowns and bridges, loose and lost teeth, and periodontal bone breakdown. These diagrams and descriptions were compared, tooth by tooth, to the Lehrter Railroad station skeletal remains when oral pathology expert Reidar Fauske Sognnaes was given permission to examine them. His conclusion from the comparison was that one of the skulls was Bormann's and that he had indeed died in that area in 1945.[7]

However, rumors that Bormann had fled Berlin persisted, and so the German general public prosecutor and Bormann's family decided to ask for the identification of the skeletal remains through the analysis of DNA. To achieve this, 2-cm-thick bone slices were removed from the middle of two bones (the right femur and the left tibia), and a DNA extraction procedure was performed. The researchers decided at the time to analyze mtDNA because they considered it better than autosomal DNA for the identification of the remains (due to the lack of recombination, its existence in multiple copies, and its resistance to degradation). For this reason, they also extracted DNA from the blood sample collected from an 83-year-old maternal, female cousin of Martin Bormann. The mtDNA of Bormann and his cousin was expected to be the same as they had both inherited it from their maternal grandmother via their mothers. The comparisons made involved the sequences of the two hypervariable regions (HVR1 and HVR2) of the bone and blood samples, as well as a reference sequence. To ensure consistency, all mtDNA analyses were performed by two different persons in two different laboratories. The HVR1 and HVR2 sequences of the mtDNA obtained from

---

[7] Sognnaes, R. F. (1976). Talking teeth: The developing field of forensic dentistry can increasingly aid the legal and medical professions in problems of identification. *American Scientist, 64*(4), 369–373; see also Sognnaes, R. F. (1973). Dental identification of Hitler's deputy Martin Bormann. *The Journal of the American Dental Association, 86*(2), 305–310.

USING DNA ANCESTRY EVIDENCE TO RETRACE HISTORY 95

the skeletal remains found near the Lehrter station were identical to the respective mtDNA sequences of Martin Bormann's cousin. The researchers noted that they did not find that particular sequence elsewhere. Given the available evidence, their conclusion was that the skeletal remains were those of Bormann.[8]

In this case, the combination of mtDNA analysis with genealogical information—the latter absolutely necessary to find Bormann's maternal cousin and compare his DNA to hers—was sufficient to establish that Martin Bormann had indeed died in Berlin in 1945 and did not flee to South America, as rumors had it.

President Jefferson and Sally Hemings's Children

Thomas Jefferson was one of the Founding Fathers of the United States, the principal author of the Declaration of Independence (1776) while he was a member of Congress, and the third president of the United States (1801–1809).[9] After 1802, Jefferson had to deal with rumors about having a Black enslaved person as mistress, who had also borne him several children. The name of this woman was Sally Hemings. Her mother, Betty, was enslaved by John Wayles, whose daughter, Martha, had married Jefferson the previous year. As Sally had a lighter skin color than her mother, there were also rumors that Wayles was her father, and therefore that Sally and Martha were half-sisters. Wayles died soon after Sally's birth, and she and her family were acquired by Jefferson and went to live in Monticello, Virginia. After the death of his wife, Martha, a few years later, Jefferson accepted an appointment in France, where Sally also went. After a short stay in France, they all returned to Virginia. But as Jefferson was appointed secretary of state under President Washington, he left Virginia and he was generally away for the subsequent 12 years, during which he also became vice president and then president. In the meantime, Sally gave birth to several children, some of whom resembled Jefferson. At least four of these children were said to have been fathered by Jefferson: Beverley, born in 1798; Harriet, born in 1801; Madison, born in 1805; and Eston, born in 1808. Many people argued that because Jefferson

---

[8] Anslinger, K., Weichhold, G., Keil, W., Bayer, B., and Eisenmenger, W. (2001). Identification of the skeletal remains of Martin Bormann by mtDNA analysis. *International Journal of Legal Medicine*, *114*(3), 194–196.

[9] https://www.whitehouse.gov/about-the-white-house/presidents/thomas-jefferson/

96 ANCESTRY REIMAGINED

had been away for most of the time during those years, he could not have been the father of these children. However, if one considers when exactly these children were born, it seems that Jefferson was in Monticello when Sally should have become pregnant.[10]

In 1998, a DNA study was published that compared the YDNA sequences of the descendants of Field Jefferson, Thomas Jefferson's paternal uncle, to those of the descendants of Thomas Woodson and of Eston Hemings, Sally Hemings's first and last son, respectively. The descendants of both Thomas and Eston believed that Thomas Jefferson had been their father. However, there was no documentation in support of this, and there were even rumors that Eston's father was either Samuel or Peter Carr, who were the sons of Jefferson's sister Martha. In order to provide evidence for or against these beliefs, the researchers analyzed 19 polymorphic YDNA markers from the following:

- five male-line descendants of two sons of Thomas Jefferson's paternal uncle, Field Jefferson
- five male-line descendants of two sons of Thomas Woodson
- one male-line descendant of Eston Hemings
- three male-line descendants of three sons of John Carr, grandfather of Samuel and Peter Carr

What was found was an exact match for all markers between the descendants of Eston Hemings and four of the five descendants of Field Jefferson. In contrast, there were many differences between the descendants of Eston Hemings and the descendants of both Thomas Woodson and John Carr. Particularly telling is that the researchers detected a specific set of DNA variations so rare that they had never been found in anyone outside the Jefferson family. With little historical evidence that another male-line descendant of Field Jefferson and relative of Thomas Jefferson could have also been the father of Eston Hemings, these findings were taken to suggest that Thomas Jefferson was the father of Eston Hemings, but not of Thomas Woodson.[11]

---

[10] Graham, P. M. (1961). Thomas Jefferson and Sally Hemings. *The Journal of Negro History, 46*(2), 89–103; see also Ellis, J.J. (2000). Jefferson: Post-DNA. *The William and Mary Quarterly, 57*(1), 125–138; Nicolaisen, P. (2003) Thomas Jefferson, Sally Hemings, and the question of race: An ongoing debate. *Journal of American Studies, 37*(1), 99–118.

[11] Foster, E. A. et al. (1998). Jefferson fathered slave's last child. *Nature, 396*, 27–28.

USING DNA ANCESTRY EVIDENCE TO RETRACE HISTORY   97

The conclusions became widely known. Daniel P. Jordan, president of the Thomas Jefferson Memorial Foundation, Inc., published a statement in which he concurred with the committee's findings. [12] However, not everyone was convinced. Another group of experts who studied the evidence arrived at the conclusion that a group of approximately 25 known Virginia men believed to carry the same Y chromosome with Thomas Jefferson, including his nephews and cousins as well as his younger brother Randolph, could have been the father of Sally Hemings's children. According to them, despite the evidence that Jefferson was in Monticello when most of Sally Hemings's children should have been conceived, the allegation was in no way proven and the case was not yet closed. [13] This is a key point, worth considering further. The fact that the descendants of Field Jefferson and Eston Hemings have exactly the same YDNA markers does not prove in any way that Thomas Jefferson was Eston's father, because several other male relatives of Thomas Jefferson would also have these very same markers and so could have also fathered Eston. Therefore, Thomas Jefferson could have been Eston's father, but there is no way to confirm that he indeed was.

Thus, DNA provides crucial, but insufficient evidence to close this case. This is why other kinds of evidence, such as historical or genealogical evidence, are required, and this is even more clearly shown in the next story.

The Romanovs

The Romanovs were the last royal family of Russia that ruled the country since 1613. Tsar Nicholas II was the last Romanov emperor, from 1894 until 1917, when he was forced to abdicate his throne in the aftermath of the Bolshevik Revolution. His wife, Alexandra, was originally a German princess and a granddaughter of Queen Victoria of England. They had five children: four daughters, Olga, Tatiana, Maria, and Anastasia, and one son, Alexei. On the night of July 16, 1918, the family and the four members of their staff were executed in the basement of the building in which they

---

[12] Jordan, D. P. (2000). Statement on the TJMF Research Committee Report on Thomas Jefferson and Sally Hemings. *Report of the Research Committee on Thomas Jefferson and Sally Hemings.* https://www.monticello.org/thomas-jefferson/jefferson-slavery/thomas-jefferson-and-sally-hemings-a-brief-account/research-report-on-jefferson-and-hemings/statement-on-the-report-by-tjmf-presid ent-daniel-p-jordan/

[13] Turner, R. F. (Ed.) (2011). *The Jefferson-Hemings controversy: Report of the Scholars Commission.* Durham, NC: Carolina Academic Press.

98 ANCESTRY REIMAGINED

were held in Ekaterinburg. The remains of the Romanovs were discovered in a mass grave in 1979, but this was kept secret until 1991. The announcement by two amateur investigators, Gely Ryabov and Alexander Avdonin, that they had discovered a mass grave closer to Ekaterinburg initiated an official investigation by the Russian Federation. The remains found were badly damaged, having bullet wounds and bayonet marks. Nine distinct corpses were identified. As there was no conclusive evidence about their identity, the authorities turned to DNA analyses.[14]

The researchers identified four male and five female bodies, and found patterns in DNA that would be expected to exist in a family for five of the nine bodies: two parents and three children. To establish that this family was indeed the Romanovs, the researchers analyzed the hypervariable regions (HVR1 and HVR2) in their mtDNA. The sequences of the individuals presumed to be Alexandra and her children were identical, as one would expect for a mother and her children. These mtDNA sequences were also found to be identical to those of Prince Philip, the Duke of Edinburgh and husband of Queen Elizabeth II, who was a grand-nephew of continuous maternal descent from Alexandra. This confirmed the identity of Alexandra and her children. The mtDNA sequence of the presumed Tsar Nicholas was compared to those of two relatives of continuous maternal descent from his maternal grandmother Louise of Hesse-Cassel, queen of Denmark. Again, the mtDNA sequences were found to be the same, with the exception of a single nucleotide (in the position 16169). Further examination of this sequence showed that it was a case of heteroplasmy (see Box 5.1). Nicholas was found to exhibit two kinds of mtDNA sequences, one with a T in position 16169 that the descendants of his maternal grandmother also had, and one with a C in that position, at a ratio of approximately 1: 3.4, respectively. From these findings, it was also concluded that the body of Alexei and one of his sisters was missing.[15]

The heteroplasmy found in the remains of Nicholas II was a concern, so it was crucial to find further evidence supporting the authenticity of these remains. To achieve this, the remains of Tsar's Nicholas II brother, Grand Duke of Russia Georgij Romanov who had died in 1899, were exhumed from St. Peter and Paul Cathedral in St. Petersburg. The mtDNA (both HVR1 and

---

[14] https://www.nationalgeographic.com/history/world-history-magazine/article/romanov-dynasty-assassination-russia-history

[15] Gill, P., Ivanov, P. L., Kimpton, C., Piercy, R., Benson, N., Tully, G., and Sullivan, K. (1994). Identification of the remains of the Romanov family by DNA analysis. *Nature Genetics*, 6(2),130–135.

USING DNA ANCESTRY EVIDENCE TO RETRACE HISTORY 99

Box 5.1 What Is Heteroplasmy?

---

mtDNA mutates faster than autosomal DNA.[1] However, mutations do not necessarily cause problems. One reason seems to be the fact that multiple copies of mtDNA molecules coexist, so that any problem caused by a mutation in one of them may be compensated by other unchanged mtDNA molecules. The phenomenon in which individuals have mtDNA molecules with different DNA sequences, because of the coexistence of mtDNA molecules that underwent mutation and others that did not, is called heteroplasmy (if all mtDNA molecules within a cell or organism are identical, the phenomenon is called homoplasmy). The level of heteroplasmy can vary between cells, tissues, or individuals of the same family. It seems that most humans have multiple mtDNA molecules, and so heteroplasmy is very common, even though only a small proportion of mtDNAs are passed from mother to offspring.[2] Recent studies have shown that offspring can have heteroplasmic variants that are not present in their mother, and that must have emerged anew.[3]

[1] The estimated mutation rate for mtDNA is 200×10^{-9} mutations per base-pair per year. See Scally, A., and Durbin, R. (2012). Revising the human mutation rate: implications for understanding human evolution. *Nature Reviews Genetics, 13*(10), 745–753.

[2] This results in the lack of high-level heteroplasmy (present at levels greater than 10%) in most humans. But low-level heteroplasmy is very common in humans, and heteroplasmy at very low levels (0.5%–1%) seems to be almost universal. See Stewart, J. B., and Chinnery, P. F. (2021). Extreme heterogeneity of human mitochondrial DNA from organelles to populations. *Nature Reviews Genetics, 22*, 106–118.

[3] Wei, W., Tuna, S., Keogh, M. J., Smith, K. R., Aitman, T. J., Beales, P. L., Bennett, D. L., Gale, D. P., Bitner-Glindzicz, M. A., Black, G. C., and Brennan, P. (2019). Germline selection shapes human mitochondrial DNA diversity. *Science, 364*(6442); Zaidi, A. A., Wilton, P. R., Su, M. S. W., Paul, I. M., Arbeithuber, B., Anthony, K., et al. (2019). Bottleneck and selection in the germline and maternal age influence transmission of mitochondrial DNA in human pedigrees. *Proceedings of the National Academy of Sciences, 116*(50), 25172–25178.

HVR2) of Georgij was compared to the mtDNA of the presumed remains of Nicholas II, as well as to that of Countess Xenia Cheremeteff-Sfiri, a living maternal relative. They found that the sequence of Georgij matched those of Nicholas II, including the C/T heteroplasmy at position 16169. However, the ratio of C/T was different between the two brothers (Georgij had more T than C, whereas Nicholas II had more C than T). The heteroplasmy was the only difference between the mtDNA of Georgij and Countess Xenia; otherwise their mtDNA sequences matched each other. The researchers suggested that the heteroplasmy could have passed from the tsar's mother

100 ANCESTRY REIMAGINED

Maria Feodorovna to her sons Georgij and Nicholas II, but this did not occur in other maternal relatives. The conclusion therefore was that the match of the mtDNA sequences of Georgij Romanov and of the presumed remains of the Tsar Nicholas II, despite the heteroplasmy at the same position, was very strong evidence that the remains were indeed those of the latter.[16]

But there was still an unanswered question: what had happened to Alexei and one of his sisters? There were actually legends that some of the children might have survived. However, the remains of two burned human skeletons were found in the area near Yekaterinburg in 2007. The researchers also managed to extract and sequence complete mtDNA from the bones in the second grave and from the bones of Alexandra. Then they compared these sequences not only to one another but also to those from two different descendants of the maternal lineage of Queen Victoria, resulting in a perfect match. Furthermore, YDNA analysis showed a match not only between the remains of Nicholas II and his son Alexei, as one would expect, but also between both of them and living cousins of Nicholas II derived from continuous paternal lineages of Emperor Nicholas I. In both cases, the sequences were not found in any databases, and so it was concluded that they were restricted to these families. Finally, perhaps the most intriguing feature of this study was the extraction and analysis of DNA from traces of his blood on an excellently preserved shirt that Nicholas II wore during an assassination attempt in Japan in 1891. All analyses confirmed that the person from which the samples came was Nicholas II, as even the 16169 C/T heteroplasmy was found in the blood specimens.[17]

Another study confirmed all previous findings. The only question that was not possible to answer was which of the Romanov daughters remains, Maria's (according to Russian experts) or Anastasia's (according to US experts), were found with Alexei in the second grave. This was due to the absence of a definitive DNA reference for each sister, in the way that YDNA was used to identify Alexei, as he was the only son of Nicholas II and Alexandra. If they

---

[16] Ivanov, P. L., Wadhams, M. J., Roby, R. K., Holland, M. M., Weedn, V. W., Parsons, T. J., Ivanov, P. L., Wadhams, M. J., Roby, R. K., Holland, M. M., Weedn, V. W., and Parsons, T. J. (1996). Mitochondrial DNA sequence heteroplasmy in the Grand Duke of Russia Georgij Romanov establishes the authenticity of the remains of Tsar Nicholas II. *Nature Genetics, 12*(12), 417–420. For an overview of the debates, see Coble, M. D. (2011). The identification of the Romanovs: Can we (finally) put the controversies to rest? *Investigative Genetics, 2*(1), 1–7.

[17] Rogaev, E. I., Grigorenko, A. P., Moliaka, Y. K., Faskhutdinova, G., Goltsov, A., Lahti, A., et al. (2009). Genomic identification in the historical case of the Nicholas II royal family. *Proceedings of the National Academy of Sciences, 106*(13), 5258–5263.

USING DNA ANCESTRY EVIDENCE TO RETRACE HISTORY 101

had had another son, the identification of Alexei would not have been that conclusive.[18]

Summing Up

What the studies and the findings considered in the present chapter show is that DNA data can be used to identify individuals with a high level of accuracy. However, we can now reconsider the two points that I suggested we must keep in mind at the beginning of the present chapter.

The first one is that whereas some questions can be answered with a high degree of certainty, such as that Eston Hemings was a male-line descendant of the Jefferson family or that the family found in the two graves in Yekaterinburg were the Romanovs, other questions may remain unresolved, such as whether it was Thomas Jefferson or one of his male relatives who was Eston's father or whether it was Maria's or Anastasia's remains that were found in the second grave with Alexei's. In short, DNA data can be used to answer some questions, but not all of them.

The second point is that all interpretations of DNA data rely on evidence from additional sources. It would be impossible to draw the aforementioned conclusions if it was not for the relevant historical and genealogical information that led to the living descendants and relatives of the historical figures discussed in this chapter. DNA can be used as data only when we can compare different DNA sequences, but we need to know which sequences we should compare. This can only be decided if we have sufficient historical and genealogical information. Nevertheless, the study of DNA can be quite informative when we focus on individuals.

By now it should be clear what ancestry is. The study of genetic ancestry combined with knowledge about genealogical ancestry from other sources can provide answers to many questions. However, we often misperceive and misunderstand ancestry, resulting in particular paradoxes to which we now turn.

[18] Coble, M. D., Loreille, O. M., Wadhams, M. J., Edson, S. M., Maynard, K., Meyer, C. E., et al. (2009). Mystery solved: The identification of the two missing Romanov children using DNA analysis. *PLoS One, 4*(3), e4838.

6

We Are All Africans, Ultimately

The Search for Deep "Roots"

As I explained in Chapter 4, while the number of our genealogical ancestors increases as we go back in time, the number of those who are also our genetic ancestors decreases. I also explained that our family trees become more and more intertwined as we go back in time, and so any two people in the world will be more likely to have a common ancestor the further back we go in time. Finally, I also cited studies that have estimated that the common ancestor of all living humans must have lived just a few thousand years ago, and that every person who has lived 1,000 years ago and has had descendants would be an ancestor of every European living today. However, it seems natural to people to look for "roots" and "pure" identities. As we saw in the Preface, before taking the test Karen thought of herself as 100% East-African and Jay thought of himself as 100% English. Similar was the case for other participants in *The DNA Journey* show. For instance, Aurelie mentioned that all her family was from France;[1] Carlos said that his family was 100% Cuban;[2] and so on. As we saw in Chapter 1, this is often the case for DNA ancestry test-takers. What I described in the Preface as the genealogical imagination is how we imagine where we come from, which in turn defines who we are. It seems that we imagine that there is a single, fixed starting point in time where the beginning of a lineage leading to ourselves is found. In the present chapter I explain that it is a lot more complicated than that.

Another participant in *The DNA Journey* show was Ellaha, who said that she was originally from Kurdistan, in Iran, but arrived as a political refugee to Denmark with her siblings and parents when she was six years old. As Ellaha pointed out about Kurdish people, "We don't have a country anymore! We haven't had one for a long time. And when you are a person like me, you don't even know exactly what your grandparents are, if you look at blood. But we

---

[1] https://www.youtube.com/watch?v=mer2HG9dSdU (accessed March 1, 2022).
[2] https://www.youtube.com/watch?v=EYnutf0rqeY (accessed March 1, 2022).

Ancestry Reimagined. Kostas Kampourakis, Oxford University Press. © Oxford University Press 2023.
DOI: 10.1093/oso/9780197656341.003.0006

WE ARE ALL AFRICANS, ULTIMATELY   103

are confident enough to say that we are a people of our own." The results that Ellaha received stated that she was 79% from Iran and Caucasus, whereas she also had some European Jewish ancestry. The test results also indicated that another participant, Waj, was a distant cousin of hers. "We are 99.96% sure" the interviewers told her "that you guys shared a common ancestor, somewhere between 150 and 225 years ago." The show ended with Ellaha meeting her distant cousin Waj.[3] Touching for sure, but there is a lot more to consider in this case.

Recall that in Chapter 4 we assumed a generation time of 30 years. To simplify the calculations, let us assume here a generation time of 25 years. Therefore, Ellaha and Waj had a common ancestor between six and nine generations ago. This was a common genetic ancestor; both Ellaha and Waj had inherited DNA segments from that person, which resulted in a match when their DNA was analyzed. But this was also a common genealogical ancestor among a total number that would (theoretically) range between 64 (2^6) six generations ago and 512 (2^9) nine generations ago for each one of them. The analysis of DNA somehow privileges this common ancestor, because Ellaha and Waj happened to inherit common DNA segments from that person. But does this mean that all the other genealogical ancestors are irrelevant? To visualize this, take a look again at Figure 4.3. George and his female cousin share an identical DNA segment that they have both inherited from their (maternal and paternal, respectively) grandfather, who is one of their most recent common ancestors (the other is his wife, their grandmother). Should we privilege their grandfather over their grandmother just because it is the former and not the latter for whom genetic ancestry was established?

I think not, but this is exactly what DNA ancestry testing does. Once we find DNA matches, we point to the common genetic ancestor from whom the respective DNA segment was inherited, and therefore to that person's presumed or inferred geographical region of origin, as if there were no other genealogical ancestors. The problem here is the resulting misrepresentation and misperception of our past. Whereas our past and our ancestors would be best represented by a complex network of family trees that would overlap more and more the further we went back in time, we privilege particular paths within those family trees just because those can be inferred from the similarities with other people that we find today in our DNA. Thus, a complex past with numerous genealogical ancestors becomes oversimplified

[3] https://www.youtube.com/watch?v=RATWGJbGDkA (accessed March 1, 2022).

104   ANCESTRY REIMAGINED

because we tend to privilege those to whom we find ourselves genetically connected, by relying on the few of its traces still left in our DNA.

Whereas our "roots" expand continuously as we go back to the past, we selectively narrow them down based on the traces that we find on our DNA. What this can provide us with eventually is a fixed starting point (in time and place) to which we can locate our ancestry. This way of thinking is best exemplified in the studies searching for a genetic Eve and Adam. The first studies of human DNA variation were based on mtDNA,[4] which attracted attention in 1987, with the publication of a paper titled "Mitochondrial DNA and Human Evolution" by molecular human evolution pioneer Allan Wilson and his colleagues Rebecca Cann and Mark Stoneking.[5] The analysis of mtDNA from 147 individuals coming from five different populations supported several interesting conclusions. The first one was that humans had evolved in Africa and had dispersed out of it, probably multiple times. The second one was that there existed a common female ancestor of all the people tested, from whom their mtDNA derived. This female ancestor was estimated to have lived in Africa about 200,000 years ago, and became known as the "mitochondrial Eve" or "African Eve" (hereafter mt-Eve). It must be noted that even though the authors of the article themselves wrote in the abstract that "All these mitochondrial DNAs stem from one woman who is postulated to have lived about 200,000 years ago, probably in Africa," what they found was a common ancestral mtDNA sequence, rather than a person who was the common ancestor. The genealogical tree they produced represented different types of mtDNA, which had a coalescence time in the past. Coalescence time is the time required, going backward in time, for two or more lineages to coalesce—that is, to join one another—into their ancestral lineage. This point in time was what the mt-Eve really represented.

What the methods used by the researchers showed was that it was possible to arrange the various mtDNA sequences in clusters on the basis of their similarities and differences. Because mtDNA does not undergo recombination and because mutations occur to it, it is possible to compare sequences

[4] The main reason for the initial use of mtDNA was that it existed in multiple copies in cells—which was crucially important in the early 1980s before PCR (polymerase chain reaction, a reaction that can amplify DNA molecules and thus produce multiple copies of the same DNA segment for analysis, even from a single cell) was invented—and because it is much shorter than autosomal DNA. But even after the use of PCR became widespread, the analysis of mtDNA in evolutionary studies remained important because—as we saw in Chapter 4—mtDNA is transmitted mostly unchanged across generations.

[5] Cann, R. L., Stoneking, M., and Wilson, A. C. (1987). Mitochondrial DNA and human evolution. *Nature*, 325(6099), 31–36.

WE ARE ALL AFRICANS, ULTIMATELY   105

and infer, based on their similarities and differences, which is the oldest one from which the others could have been derived. The idea behind all this was that all humans could trace their ancestry through their matrilineal lineages (from one's mother, to one's maternal grandmother, and so on). The oldest sequence could be supposed to belong to the female common ancestor. But even if mt-Eve had actually existed, this does not mean that she was the first or the only woman to ever have lived. Rather, several other women could have existed before or at the same time with her. What the findings of mtDNA supported was the conclusion that, if such a person actually existed, it was only she who had had descendants by a direct female line in today's human populations. Her contemporaries did not manage this, either because they had no children, or because, at some point in time, they only had male descendants who did not pass on their mtDNA to their own descendants.

This finding had important implications for our understanding of human evolution. During the 20th century, different theories of human evolution were proposed. One was that modern humans evolved in Africa and then dispersed around the world, known as the "out of Africa" theory. Another was that modern humans evolved multiple times at various places in the world, which is known as the "multiregional" theory. The "discovery" of mt-Eve was crucial for supporting the "out of Africa" theory of human evolution. If the "multiregional theory" were correct, one would have expected to find mtDNA sequences among people living today that pointed to a common ancestor who had lived around 2 million years ago, when *Homo erectus* began their dispersal around the globe. In such a case, there should have been a continuity between the various populations of *H. erectus* in different regions of the world and their current descendants there.[6] It is interesting to point out that in his book *The Descent of Man*, Charles Darwin had already speculated that Africa was the likely place of origin of humans: "It is therefore probable that Africa was formerly inhabited by extinct apes closely allied to the gorilla and chimpanzee; and as these two species are now man's nearest allies, it is somewhat more probable that our early progenitors lived on the African continent than elsewhere."[7]

In 1991, the remains of a human, the so-called Iceman, were found in the Italian Alps. The dating of the remains indicated that that person had lived

---

[6] Stringer, C. (2012). *Lone survivors: How we came to be the only humans on Earth*. New York: St Martin's Griffin, pp. 18–26.
[7] Darwin, C. R. (1871). *The descent of man, and selection in relation to sex* (Vol. 1, 1st ed.). London: John Murray, p. 199 (available at http://darwin-online.org.uk)

106 ANCESTRY REIMAGINED

between 5,000 and 5,350 years ago. As the body had been deep frozen for a long time, its tissues had been preserved and so it was possible to extract and analyze that person's DNA. The analysis showed it to be similar to the DNA of present-day Europeans. Subsequent analyses of mtDNA showed that more than 95% of modern Europeans belonged to one of seven mtDNA sequence clusters. This conclusion was transformed in a popular book by geneticist Bryan Sykes to the claim that almost every person of European descent living today could trace their ancestry back to one of seven women: the seven daughters of Eve, to whom Sykes gave the names Ursula, Tara, Helena, Katrine, Xenia, Jasmine, and Valda.[8] The problem with representations such as these is that an initially statistical estimate based on molecular data was transformed to hypothesized historical persons—Eve and her seven "daughters"—who were considered as the ancestors of present-day humans. This masks the fact that different methods and assumptions might have produced different estimates, and therefore a different story.[9]

The initial discussions about mt-Eve contained speculations about the existence of a genetic Adam. Not surprisingly, such a "discovery" followed a few years later by geneticist Michael Hammer. The first study, based on the analysis of a small segment of the Y chromosomes of 15 humans (eight Africans, two Australians, three Japanese, two Europeans) and 4 chimpanzees, estimated the existence of a common ancestral human Y chromosome around 188,000 years ago. Like the 1987 paper on mtDNA, the 1995 paper on YDNA also questioned the multiregional hypothesis of human evolution.[10] A subsequent study compared YDNA variation to mtDNA variation, and the authors concluded that they had different coalescence times. Their estimate for the coalescence time of the mtDNA sample they used was between 120,000 and 474,000 years, whereas their estimate for the coalescence time of the YDNA sample they used was between 37,000 and 49,000 years.[11] More

---

[8] Sykes, B. (2002). *The seven daughters of Eve: The astonishing story that reveals how each of us can trace our genetic ancestors.* London: Corgi Books. Sykes founded in 2000 the company Oxford Ancestors, a spin-off from the University of Oxford, one of the first companies to provide direct-to-consumer genetic testing for ancestry. The relationship between the academic science that develops the methods for ancestry testing and the corporate science that sells the tests to consumers is an important one, but it would require a book of its own and falls outside the scope of the present book. See, for instance, Bolnick, D. A., Fullwiley, D., Duster, T., Cooper, R. S., Fujimura, J. H., Kahn, et al. (2007). The science and business of genetic ancestry testing. *Science, 318*(5849), 399–400.

[9] Oikkonen, V. (2018). *Population genetics and belonging: A cultural analysis of genetic ancestry.* Cham, Switzerland: Palgrave Macmillan, pp. 47–48.

[10] Hammer, M. F. (1995). A recent common ancestry for human Y chromosomes. *Nature, 378*(6555), 376–378.

[11] Whitfield, L. S., Sulston, J. E., and Goodfellow, P. N. (1995). Sequence variation of the human Y chromosome. *Nature, 378*(6555), 379–380.

WE ARE ALL AFRICANS, ULTIMATELY 107

recent studies have reached different conclusions. For instance, in order to overcome the limitations of previous studies, researchers sequenced the complete Y chromosomes and the mtDNA of 69 males, as well as the mtDNA of 24 females, from nine populations.[12] Based on these data, they concluded that the most recent ancestor for the Y chromosome must have lived between 120 and 156 thousand years ago, whereas the most recent ancestor for mtDNA must have lived between 99 and 148 thousand years ago. This finding put the Y-Adam not only closer in time to the mt-Eve but also closer to nowadays.[13] But, again, we should keep in mind that these are estimates. For instance, another more recent study of 456 Y chromosome sequences from various places in the world dated the Y-chromosomal most recent common ancestor in Africa between 192 and 307 thousand years ago.[14]

But why do the estimates of the various studies differ from one another? The reason is that these estimates are the outcome of calculations. As such, they depend on several variables, most importantly the methods for calculating the coalescence time for each lineage, as well as on which human samples the DNA sequences come from. Different methods and different samples can yield different estimates, and in turn conclusions. Because mt-Eve and Y-Adam are the outcome of DNA sequence data and statistical analysis, the idea that the DNA of all humans might come from these two ancestors is erroneous. Even if one believes that the Eve and Adam of the scriptures were our genealogical ancestors, they cannot be our genetic ancestors for the reasons already explained in Chapter 4.[15]

What does all this have to do with DNA ancestry testing? The studies considered in this section attempted to figure out and describe the genetic history of the human species. Genetic history is the study of the origins of a population and its evolutionary relations to other populations. Its aim is to compare the genetic diversity of human groups that are distinct from one

[12] These were San Bushmen from Namibia, Mbuti Pygmies from the Democratic Republic of Congo, Baka Pygmies and Nzebi from Gabon, Mozabite Berbers from Algeria, Pashtuns from Pakistan, Cambodians, Yakut from Siberia, and Mayans from Mexico.

[13] Poznik, G. D., Henn, B. M., Yee, M. C., Sliwerska, E., Euskirchen, G. M., Lin, A. A., et al. (2013). Sequencing Y chromosomes resolves discrepancy in time to common ancestor of males versus females. *Science, 341*(6145), 562–565.

[14] Karmin, M., Saag, L., Vicente, M., Sayres, M. A. W., Järve, M., Talas, U. G., et al. (2015). A recent bottleneck of Y chromosome diversity coincides with a global change in culture. *Genome Research, 25*(4), 459–466.

[15] See also Swamidass, S. J. (2019). *The genealogical Adam and Eve: The surprising science of universal ancestry*. Westmont, IL: InterVarsity Press.

108 ANCESTRY REIMAGINED

another culturally or politically, such as nations or ethnic groups.[16] DNA ancestry tests offer "personalized genetic histories."[17] What this entails is that in order to figure out the personalized genetic histories of individuals, we need to rely on the genetic histories of populations as a point of reference. Therefore, in order to understand what DNA ancestry tests can and cannot "tell" us, we need to understand how scientists study the DNA diversity of populations and what inferences can be made about their genetic histories. The key idea here is that in order for DNA ancestry testing to offer ethnicity estimates and the reconstruction of a person's genetic ancestry, it has to rely on the study of human DNA variation at the population level.

Comparing the DNA Variation of Human Populations

The study of DNA variation of contemporary human populations can provide valuable insights about how they may be related to one another, as well as about historical events such as migrations that in turn may explain the differences among them. Individuals are compared to one another on the basis of which combinations of DNA markers they have. Because the results depend on which DNA markers have been selected for the comparison, a careful selection has to be made and interpretations should be made with caution. Given that we can estimate how often human DNA undergoes mutation,[18] from the number of differences we find among different DNA sequences we can estimate the time of their divergence from a common ancestor. The main idea is that the more differences we find between two DNA sequences, the further back in time their common ancestor has existed. The reason for this is that the more time has passed since their divergence, the more mutations are likely to have occurred and the more differences are likely to be found. I must note, though, that the outcome of such calculations

[16] Abu El-Haj, N. A. (2012). *The genealogical science: The search for Jewish origins and the politics of epistemology.* Chicago: University of Chicago Press, pp. 3, 14.

[17] Shriver, M. D., and Kittles, R. A. (2004). Genetic ancestry and the search for personalized genetic histories. *Nature Reviews Genetics, 5*(8), 611–618; Shriver, M. D., and Kittles, R. A. (2008). Genetic ancestry and the search for personalized genetic histories. In B. Koenig, S. Lee, and S. Richardson (Eds.), *Revisiting race in a genomic age* (pp. 201–214). New Brunswick, NJ: Rutgers University Press.

[18] The mutation rate for humans during the past million years has been proposed to be around 0.5 × 10^{-9} mutations per base-pair per year. Scally, A., and Durbin, R. (2012). Revising the human mutation rate: Implications for understanding human evolution. *Nature Reviews Genetics, 13*(10), 745–753; Scally, A. (2016). The mutation rate in human evolution and demographic inference. *Current Opinion in Genetics & Development, 41*, 36–43.

WE ARE ALL AFRICANS, ULTIMATELY 109

is only an estimate that depends on the statistical methods and the particular assumptions and parameters of the models used.[19]

Several human genome projects have aimed to study human DNA variation.[20] The most widely known ones are the Human Genome Diversity Project (HGDP),[21] the Genographic Project (GP),[22] the Haplotype Map (HapMap) project,[23] the 1,000 Genomes Project (1KGP),[24] and the Simons Genome Diversity Project (SGDP).[25] A detailed discussion of these projects falls outside of the scope of the present book, especially as several detailed, critical analyses have already been written for some of them.[26] However, their data have been used in many subsequent studies and so it is necessary to be aware of their main features, which are summarized in Table 6.1. It should be noted that these projects began with different assumptions. For instance, the HapMap Project intended to compare what the researchers considered as the main continental groups, the HGDP intended to study what were considered as Indigenous, and in this sense "pure," groups, whereas the other two projects looked at human DNA diversity more broadly.

Overall, the most significant conclusion of these projects has been that human genomes are very similar. For instance, an important finding of the 1KGP was that a typical human genome differs from the reference human

[19] Zhou, J., and Teo, Y. Y. (2016). Estimating time to the most recent common ancestor (TMRCA): Comparison and application of eight methods. *European Journal of Human Genetics, 24*(8), 1195–1201.

[20] These are different from the Human Genome Project (HGP) that aimed to find the sequence of the human genome. The HGP begun in the early 1990s and was officially concluded in 2003, with an announcement of initial human DNA sequences in 2000. I write initial, because it was only recently, in 2021, that the "complete" sequence of the human genome was announced, even though the DNA of the Y chromosome is not yet completely sequenced. Reardon, S. (2021). A complete human genome sequence is close: How scientists filled in the gaps. *Nature, 594*, 158–159.

[21] Cann, H. M., De Toma, C., Cazes, L., Legrand, M. F., Morel, V., Piouffre, L., et al. (2002). A human genome diversity cell line panel. *Science, 296*(5566), 261–262; Cavalli-Sforza, L. L. (2005). The human genome diversity project: Past, present and future. *Nature Reviews Genetics, 6*(4), 333–340.

[22] Wells, S. (2007). *Deep ancestry: Inside the Genographic Project.* Washington, DC: National Geographic.

[23] The International HapMap Consortium. (2003). The International HapMap Project. *Nature, 426*, 789–796; https://www.genome.gov/10001688/international-hapmap-project

[24] The 1000 Genomes Project Consortium. (2010). A map of human genome variation from population-scale sequencing. *Nature, 467*, 1061–1073; https://www.internationalgenome.org

[25] Mallick, S., Li, H., Lipson, M. et al. (2016). The Simons Genome Diversity Project: 300 genomes from 142 diverse populations. *Nature, 538*, 201–206.

[26] Reardon, J. (2005). *Race to the finish: Identity and governance in an age of genomics.* Princeton, NJ: Princeton University Press; M'charek, A. (2005). *The Human Genome Diversity Project: An ethnography of scientific practice.* Cambridge: Cambridge University Press; Nash, C. (2015). *Genetic geographies: The trouble with ancestry.* Minneapolis: University of Minnesota Press; Sommer, M. (2016). *History within: The science, culture, and politics of bones, organisms, and molecules.* Chicago: University of Chicago Press.

110 ANCESTRY REIMAGINED

Table 6.1 Studies of Human DNA Diversity

Project	No. of Samples	No. of Populations	Choice of Populations
HGDP	1,064	52	Indigenous groups
HapMap	270[a]	5	Continental groups
1KGP	2,504 (HapMap) [b]	26	Diverse groups
SGDP	300	142	Diverse groups

[a] Ninety samples collected in 1980 by the CEPH (see Chapter 2) from a US Utah population with Northern and Western European ancestry (30 parent-child trios), and new samples collected from 90 Yoruba people in Ibadan, Nigeria (30 parent-child trios), 45 unrelated Japanese in Tokyo, Japan, and 45 unrelated Han Chinese in Beijing, China. See The International HapMap Consortium (2003). The International HapMap Project. *Nature*, *426*, 789–796.

[b] This has recently been expanded with 698 additional related samples. See Byrska-Bishop, M. et al. (2021) High coverage whole genome sequencing of the expanded 1000 Genomes Project cohort including 602 trios. BioRxiv

genome at 4.1 million to 5 million sites.[27] Given that the reference genome is approximately 3.1 billion base-pairs long,[28] this means that a typical genome is between 0.132% and 0.161% different from the reference genome. This practically means that we can assume that human genomes are about 99.9% identical. But even though the study of the human genome points unequivocally to the unity of the human species, there is a lot of interest in its very minor portion that varies among individuals. Therefore, whenever you read anything about human DNA variation and differences among human populations, you must keep in mind that these are found in a very minor portion of our DNA. If whole genomes were compared, the variation among them would be very tiny. Caution is thus required because the study of ancestry can focus on these minor differences and blind us to the overall picture. This attention to the local instead of the global can obscure our enormous genetic relatedness and blur the most important conclusion of these human genome variation projects: we are all family (see also Chapter 7).

A common assumption in all these projects is that human populations can be clearly delimited. It may sound strange, but this is far from simple and straightforward (I consider this in detail in Chapter 8). In the genetic sense, a

[27] 1000 Genomes Project Consortium. (2015). A global reference for human genetic variation. *Nature*, *526*(7571), 68–74.

[28] 3,096,649,726 base-pairs, to be precise; see https://www.ensembl.org/Homo_sapiens/Info/Annotation; Yates, A. D., Achuthan, P., Akanni, W., Allen, J., Allen, J., Alvarez-Jarreta, J. et al. (2020). Ensemble 2020. *Nucleic Acids Research*, *48*(D1), D682–D688.

population is a group of interbreeding individuals. This interbreeding results in the transmission of particular DNA variants from one generation to the next. But this does not mean that all variants are transmitted in the same manner. If we could measure the frequencies of DNA variants of a particular population at different times, we would find two main kinds of differences. First, the frequencies of the DNA variants that already exist can fluctuate across generations because some individuals but not others had offspring and thus passed on their variants to the next generation. This can happen if populations are small, if mating is not random, or if the survival and reproduction of some individuals was favored in the particular environment (natural selection). Second, new DNA variants can be introduced in the population due to mutations or migration.[29] All this entails that populations are not static across time; rather their constitution constantly changes because their members die out and are replaced by those who are born. Therefore, what we have are not the same populations across time, but consecutive populations consisting of different individuals the lifespans of whom may overlap.

Populations are defined with reference to a particular area and a particular time. For example, contemporary populations consist of particular individuals who are currently living in particular areas. From the perspective of genetics, what is of interest is both the genetic constitution of each one of the individuals of the population, and of the population as a whole. This genetic constitution can be described in various ways; for instance, we might be interested in particular single nucleotide polymorphisms (SNPs) that exist in the DNA of these individuals. Some individuals may have, say, version *A* of an SNP; others may have version *B*, *C*, or *D*. If we manage to figure out which versions of these variants the individuals of a population have, we can thereafter describe the genetic constitution of a population in terms of the frequencies of variants *A*, *B*, *C*, or *D*. This is often achieved by measuring a sample of members of a population, as it is practically impossible to study every single individual. For this reason, the sample studied has to be of sufficient large *size* in order to be *representative* of the population (the size and the representativeness of a sample are two very important variables to which I return in Chapter 8).

---

[29] Falconer, D. S. (1989). *Introduction to quantitative genetics* (3rd ed.). New York: John Wiley & Sons/Longman Scientific & Technical, pp. 4–7.

112 ANCESTRY REIMAGINED

As soon as we find the frequencies of particular variants in particular populations, we have information about their DNA variation (see also Box 2.1).[30] Thus, it is possible to compare two populations in terms of their DNA variation and make inferences about how different or similar they are. The differences between two populations might be qualitative, as two populations may differ in which variants their individuals have. For instance, the individuals of a population may only have variants A, B, and C, whereas the individuals of another population may have variants B, C, and D. The differences between two populations might also be quantitative; that is, two populations might differ in the frequencies of the variants their individuals have. For instance, the individuals of two populations may have all four variants A, or B, or C, or D, but A and B may be in relative high frequency in one population, whereas C and D are in relative high frequency in the other.

Let us consider a simple example. Imagine that an autosomal DNA variant V exists in two alternative versions, $V1$ and $V2$. Imagine also that these two variants are found in four populations P1, P2, P3, and P4. Because we have pairs of homologous chromosomes and two copies of our genome (Box 4.1), each individual will carry two versions of variant V: either $V1V1$, or $V1V2$, or $V2V2$. The exact combination of these variants that each individual has can be described as the genotype of the individual. A person who has the same variant twice is a homozygote ($V1V1$ or $V2V2$), whereas a person who has two different variants is a heterozygote ($V1V2$). Similar would be the case for another variant W, with alleles $W1$ and $W2$. If we can find sufficiently representative samples of the four populations, we can figure out which variants each individual in each population carries, and calculate their frequencies. If we also know the sizes of these populations, based on the frequencies found in the samples, we can calculate the number of people in each population who are expected to have each genotype. If we know all this, we can then calculate the frequency of each variant in each population, as shown in Table 6.2.

We can get a first, visual sense of how close genetically these four populations are by putting the frequencies of variants $V1$ and $W1$ (or any other V and W combination) on a two-dimensional plot, as in Figure 6.1.

---

[30] The amount of variation within a population is measured and expressed as its variance, which is a statistical measure of the spread between values in a data set. More specifically, variance measures how far each value in the set is from the mean and thus from every other number in the set. When values are expressed as deviations from the population mean, then the variance is the mean of the squared values.

Table 6.2 The Hypothetical Frequencies of Variants *V1*, *V2*, *W1*, *W2* in Populations P1–P4

	V1	*V2*	*W1*	*W2*
P1	50%	50%	60%	40%
P2	65%	35%	75%	25%
P3	53%	47%	50%	50%
P4	25%	75%	30%	70%

The first inference from this figure is that populations P1 and P3 are more similar to each other than what they are to populations P2 and P4, or those two to each other. This is an oversimplified representation of how population geneticists compare populations, as they use numerous variants to compare many populations. However, I believe that it gives you a sense of the kinds of comparisons made. The more similar the frequencies of particular variants in two populations are, the closer genetically these populations are considered to be. This may also mean that P1 and P3 have descended from the same ancestral population, and that their divergence from each other is more recent than the divergence of any of these, or their common ancestral population, from the divergence from P2 or P4.

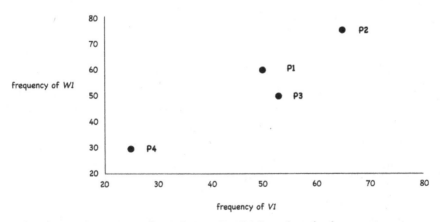

Figure 6.1 Comparison of populations P1–P4, based on the frequencies (%) of variants *V1* and *W1*. Populations P1 and P3 are considered to be closer genetically to each other compared to the others, as they have more similar frequencies.

114 ANCESTRY REIMAGINED

Two populations can begin to diverge from each other if their members do not interbreed anymore, for instance because they have become geographically separated. Until that point these populations had been part of the same population, and so they initially have similar DNA variant frequencies. The longer is the time of separation, the higher the level of divergence will be, and this will be reflected in the differences in their frequencies of DNA variants. It is important to note that the results of such comparisons depend strongly on the variants used; therefore, the more variants are used, the more authentic the resulting image is considered to be as it is assumed to cover a large amount of the extant DNA variation. Population geneticists use computer software and huge databases to make comparisons of different populations. These comparisons may involve all kinds of different populations, from small regions within a country to whole continents.

All studies of this kind point to a single unequivocal conclusion: ultimately, we all have ancestry from Africa.

We Ultimately All Have Ancestry from Africa

It is important to note that there is no specific point in place or time that we can identify as the point of origin of modern humans. The available evidence generally supports an out-of-Africa theory of human evolution with some interbreeding with other humans outside Africa. However, we can distinguish among three main phases in human evolution.

1. The first phase spans the time from 1 million years to 300 thousand years ago and concerns the divergence of the ancestors of modern humans from other archaic human groups. The ancestors of Neanderthals and Denisovans are estimated to have diverged from the ancestors of modern humans between 500 thousand and 700 thousand years ago. Many details are unclear and fossil evidence is not always in agreement with genomic evidence (also, Denisovans are currently rather poorly known from the fossil record).

2. The second phase, between 300 thousand and 60 thousand years ago, concerns the African origins of modern humans. However, although Africa seems to have been the center of the evolution of modern humans, neighboring parts of Southwest Asia seem to also have been important areas of this process.

3. Finally, the third phase between 60 thousand and 40 thousand years ago concerns the dispersion of modern humans throughout the world and their contacts with archaic groups such as the Neanderthals and the Denisovans. There exist different scenarios for the number and timing of the expansion(s) into Eurasia. However, evidence from fossils in Africa, genomic evidence of interbreeding with archaic human groups outside of Africa, and the fact that most DNA diversity outside of Africa appears to be a subset of African DNA diversity support a recent African origin.[31]

The last point is very important. That most DNA diversity outside of Africa appears to be a subset of African DNA diversity means that today's African populations exhibit more DNA diversity than today's non-African populations. The reason for this is that whenever humans dispersed out of Africa, this happened in the form of small populations that carried only some of the DNA variants found in the ancestral African population. Because African populations have existed for longer than non-African populations, they have had more time to accumulate diversity through mutations. This is best shown in some African populations, especially in East Africa from where the human dispersal seems to have taken place, which exhibit more similarities in their DNA sequences with European populations than with other African populations, as they have a more recent common ancestor with the former than with the latter. This is something very important to note because it may seem intuitive to group, based on morphological criteria, on the one hand, all Africans together and, on the other hand, all non-Africans together. However, DNA data do not support such a grouping. In particular, studies have provided evidence for the divergence between the ancestors of Khoesan-speaking San hunter-gatherers in southern Africa and other African populations more than 100 thousand years ago, as well as between central African Pygmy hunter-gatherers, such as Mbuti, and non-Pygmy populations around 60–70 thousand years ago.[32] Indeed, a recent study of whole-genome sequencing data from 92 individuals from 44 Indigenous African populations and 62 individuals from 32 west Eurasian

[31] Bergström, A., Stringer, C., Hajdinjak, M., Scerri, E. M., and Skoglund, P. (2021). Origins of modern human ancestry. *Nature, 590*(7845), 229–237.

[32] Campbell, M. C., Hirbo, J. B., Townsend, J. P., and Tishkoff, S. A. (2014). The peopling of the African continent and the diaspora into the new world. *Current Opinion in Genetics & Development, 29*, 120–132.

116 ANCESTRY REIMAGINED

populations has shown that the latter have more recent common ancestors with many African populations than these have with the San and Mbuti hunter-gatherers.[33]

This fact is very important to consider for interpreting and understanding the results of DNA ancestry testing. Whereas the tests supposedly distinguish between European and African ancestry, as we saw in Chapter 1, one may wonder: what African ancestry? The African ancestry as inferred from the DNA variation of the people who take the tests? Or the African ancestry as inferred from the enormous DNA variation that actually exists among people living there? Recent research has revealed in more detail the patterns of human DNA variation in Africa. It is becoming clear that today's genetic landscape is different than those of thousands of years ago, and that many of today's isolated populations inhabited larger regions in the past. Furthermore, there have been both periods of isolation among groups and geographic areas, and periods of genetic contact over large distances.[34] Most importantly, it remains the case that the use of ethnic categories with respect to African genomics should not be treated as unproblematic; African ethnicities, as all ethnicities, should not be assumed to be long-established, undisputed, and "real," but the outcome of historical, social, and political processes. Ethical considerations are also important: "The histories, knowledge and experiences of people in the African continent must drive scientific work that purports to be to their benefit."[35] Such studies can clarify the exact relatedness of human populations—we are all related anyway—but we should not forget that, ultimately, we all have African ancestry! This also entails that I am aware that terms such as "San" and "Pygmy" are not unproblematic, but in the present book I am using the terms used in the scientific articles I am citing to refrain from introducing any confusion.

Whereas modern humans probably dispersed out of Africa multiple times, some studies suggest that contemporary non-African populations descend from the same ancestral population and emerged during a single out-of-Africa dispersal, around 55–65 thousand years ago. Once this occurred, humans seem to have dispersed along two main ways, one toward mainland

---

[33] Fan, S., Kelly, D. E., Beltrame, M. H., Hansen, M. E., Mallick, S., Ranciaro, A., Hirbo, J., Thompson, S., Beggs, W., Nyambo, T., et al. (2019). African evolutionary history inferred from whole genome sequence data of 44 indigenous African populations. *Genome Biology, 20*(1), 82.

[34] Hollfelder, N., Breton, G., Sjödin, P., and Jakobsson, M. (2021). The deep population history in Africa. *Human Molecular Genetics, 30*(R1), R2–R10.

[35] Yéré, H. M., Machirori, M., and De Vries, J. (2022). Unpacking race and ethnicity in African genomics research. *Nature Reviews Genetics, 23*, 455–456.

WE ARE ALL AFRICANS, ULTIMATELY 117

Eurasia and another toward Australasia and New Guinea. Along the way, modern humans interbred with the Neanderthals, around 50-65 thousand years ago. The fact that all people of non-African ancestry studied so far have about 2% of DNA indicating a Neanderthal ancestry suggests both that this interbreeding occurred after modern humans left Africa and that contemporary humans descend from a single out-of-Africa dispersal. The first modern humans seem to have lived in Europe as early as 43 thousand years ago; however, the genetic contribution of these Paleolithic Europeans to people living today seems to be limited. Europe may have been recolonized by hunter-gatherers around 27–19 thousand years ago. Around 11 thousand years ago, there was an expansion of Neolithic farmers from central Anatolia (in what today is Turkey) toward Europe, reaching the Iberian Peninsula around 7 thousand years ago and Britain and Scandinavia around 6 thousand years ago. A third wave of migration occurred around 4.5 thousand years ago by herders from the Pontic-Caspian steppe (the area that is now Russia). Modern humans also expanded toward Asia, and it seems that East Asians and Western Eurasians diverged around 36–45 thousand years ago. There was also an expansion toward Oceania, which modern humans seem to have reached around 47.5–55 thousand years ago. The Denisovans are another group of archaic humans with whom modern humans seem to have interbred. This is most evident in the DNA of Melanesians, which contains around 3%–6% of Denisovan ancestry, but less evident in continental Southeast Asians, where it is only 0.1%–0.3%. Finally, modern humans seem to have reached America around 15–14 thousand years ago, with a northern-southern divergence occurring around 14–13 thousand years ago. These seem to have been the ancestors of modern Native Americans, having diverged from the Siberians around 23 thousand years ago.[36]

Overall, the available DNA data today indicate that all humans ultimately have African ancestry; however, it would be arbitrary to select a single, fixed point for our ancestry. But, in the context of DNA ancestry testing, the analysis of mtDNA and YDNA is supposed to be able to provide us with more specific information about our "deep" ancestry. The reason is that, in contrast to autosomal DNA that is diluted across generations, mtDNA and YDNA are transmitted almost intact (see Chapter 4). Let us see what this "deep" ancestry is about.

---

[36] Nielsen, R., Akey, J. M., Jakobsson, M., Pritchard, J. K., Tishkoff, S., and Willerslev, E. (2017). Tracing the peopling of the world through genomics. *Nature*, *541*(7637), 302–310.

118 ANCESTRY REIMAGINED

"Deep" Ancestry and Its Limitations

Crossing over (Box 4.2) does not occur in mtDNA and YDNA molecules. Because of this, these molecules have sets of nucleotides that are inherited together as a group because they have not been separated by crossing over. These sets of nucleotides are called "haplotypes," which actually means "haploid genotype." Whereas most of these nucleotides will not change and will be the same in all individuals of a population (described as monomorphic, because they have one form only), some can change due to mutations and so it is possible that a few sites vary among individuals of a population (described as polymorphic, because they have many forms). Thus, haplotypes are combinations of different polymorphic sites in DNA. Whereas most of their sequence is identical, there exist a few differences among individuals. We can thus distinguish among them on the basis of which haplotype they have.

These haplotypes can be further grouped into haplogroups: groups of related haplotypes that share particular nucleotides as a result of having been derived from the same common ancestor. The more time has passed since their divergence from their common ancestor, the more mutations will have occurred and the more the haplotypes of each haplogroup will vary. By comparing them we can make inferences: for instance, if two haplogroups are relatively similar, we can infer that the time of their differentiation is relatively recent and that they are more closely related to each other than to others. Haplogroups are designated by uppercase letter from A to Z (e.g., L), whereas subgroups are designated by numbers (e.g., L0, L1, etc.). The geographical distribution of haplogroups can provide some useful insights into populations' movements. Let us consider the mtDNA haplogroups as an example.

African mtDNAs belong to haplogroup L, of which L0 is the most ancient one and is found in the San people, whereas the L1 and L2 are found in the Pygmy people. The M and N haplogroups emerged from L3 in northeastern Africa and expanded to Europe and Asia. Haplogroups A, C, D, B, and X later emerged in America and B also in the Pacific, as shown in Figure 6.2a.[37] This sounds like a nice and straightforward narrative; however, it must be noted that there is no geographic region of the world in which we find a single or a few haplogroups only, as implied by Figure 6.2a that is based on data from

---

[37] Wallace, D. C. (2015). Mitochondrial DNA variation in human radiation and disease. *Cell*, *163*(1), 33–38.

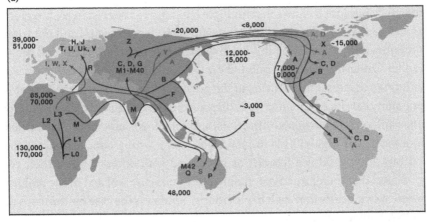

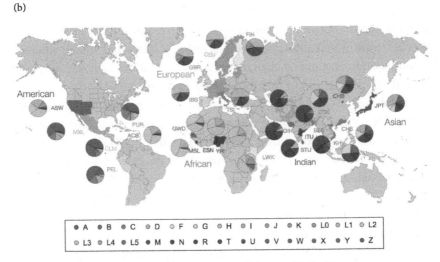

Figure 6.2 Main mtDNA haplotypes around the world. In this case, (a) represents the most frequent mtDNA haplogroups, from which inferences about past population movements are possible. (Reprinted with permission from Wallace, D. C. (2015). Mitochondrial DNA variation in human radiation and disease. *Cell*, *163*(1), 33–38.) However, (b) is a more authentic representation, as it shows that in each region multiple haplogroups coexist, as well as that samples come from specific areas in each region. (Reprinted from Rishishwar L, Jordan IK (2017) Implications of human evolution and admixture for mitochondrial replacement therapy. *BMC Genomics*, 18:140, under the terms of the Creative Commons Attribution License.)

120   ANCESTRY REIMAGINED

the 1KGP (see Table 6.1).[38] Figure 6.2b is a more accurate representation than 6.2a because it shows that multiple haplogroups coexist in each region, as well as that samples come from specific areas in each region and not from the entire regions (more on this in Chapter 8). Furthermore, as cultural geographer Catherine Nash has pointed out, Figure 6.2a, as well as my narrative of human dispersal from Africa in the previous section, creates the false impression that once human groups dispersed out of Africa, they moved only once and only to particular directions and remained static once they settled in a particular region. This in turn leads to the wrong conclusion that the various groups did not interact or interbreed with others beyond their region, and that therefore have remained in some sense isolated in the regions where we currently find each haplogroup.[39] Similar is the case for the maps of YDNA haplogroups.

One important advantage of YDNA and mtDNA is that they can provide insights about ancestry further in the past than autosomal DNA, because they are transmitted across generations rather intact—they do not undergo recombination and so any changes are due to mutations only. However, we should also keep in mind that they represent a very small portion (around 1% combined) of our DNA. This can lead to wrong conclusions. For instance, Henry Louis Gates Jr., whom we already met in Chapters 1 and 3, identifies as African American. Nevertheless, when his YDNA and mtDNA were analyzed, he came across a surprise: both his patrilineal ancestry and his matrilineal ancestry were European. As Gates himself described it: "when they analyzed my YDNA, it went to Ireland. And when they analyzed my mitochondrial DNA, it went to England. I am descended—on my father's side—from a White man who impregnated a Black woman and, on my mother's side, from a White woman who was impregnated by a Black man."[40] If Gates had only had his YDNA and his mtDNA analyzed, we would have a paradox: a person with relatively dark skin color who himself identifies as an African American would be identified by YDNA and mtDNA analyses as being Irish and English—and therefore White and European. The focus on YDNA and mtDNA would thus produce misleading results about his

---

[38] Rishishwar, L., and Jordan, I. K. (2017). Implications of human evolution and admixture for mitochondrial replacement therapy. *BMC Genomics, 18*, 140.

[39] Nash, C. (2015). *Genetic geographies: The trouble with ancestry*. Minneapolis: University of Minnesota Press, pp. 83–86, 145.

[40] https://www.npr.org/2019/01/21/686531998/historian-henry-louis-gates-jr-on-dna-testing-and-finding-his-own-roots?t=1634311634383 (accessed January 28, 2022).

ancestry, as the contributions of his other genetic ancestors to his autosomal DNA would be entirely overlooked.

Indeed, the ancestry inferences from mtDNA and YDNA can be questioned. In fact, several studies that have compared ancestry inferences based on mtDNA haplogroups and those based on autosomal SNPs have found various discrepancies. A study aimed to assess the magnitude and scope of such discrepancies by using HGDP and 1KGP data (Table 6.1).[41] The researchers classified each individual to an mtDNA haplogroup and also estimated the proportion of autosomal ancestry from each of seven continental regions.[42] Overall, each mtDNA haplogroup was found in several populations, and most populations of both the HGDP and 1KGP contained several mtDNA haplogroups. These results clearly showed that finding particular mtDNA haplogroups to be significantly associated with a specific geographical region does not entail that all people in that region will belong to the same mtDNA haplogroup. Another question of this study was whether one could predict an individual's continental ancestry from the respective haplogroup and vice versa. A key concept in this analysis is continental ancestry proportion, which is the amount of an individual's ancestry that can be considered to correspond to that individual's ancestors coming from a particular continent (discussed in detail in the Chapter 7). About half of the mtDNA haplogroups were found not to be very informative for inferring individual genetic ancestry, whereas relying on continental-ancestry proportions to predict an individual's mtDNA haplogroup was incorrect in most cases (except for haplogroup L0/L1 for the HGDP populations). Based on these findings, the researchers concluded: "Accordingly, most mtDNA haplogroups that are assigned to a continental group (e.g., an "African haplogroup" or a "European haplogroup") offer an incomplete picture of the complexity of continental ancestry within an mtDNA haplogroup. Effectively communicating this complexity to a consumer or the public poses a substantial challenge, and failure to communicate this information could perpetuate misinterpretations."[43]

Next time you order an mtDNA ancestry test, think again. What we know well enough is that all humans have an ultimate ancestry in Africa, no matter

---

[41] The study used data from 938 individuals of the HGDP and from 327 individuals from the 1KGP.

[42] These regions were sub-Saharan Africa, the Middle East, Europe, Central and South Asia, East Asia, Oceania, and the Americas.

[43] Emery, L. S., Magnaye, K. M., Bigham, A. W., Akey, J. M., and Bamshad, M. J. (2015). Estimates of continental ancestry vary widely among individuals with the same mtDNA haplogroup. *The American Journal of Human Genetics*, *96*(2), 183–193.

where our recent ancestors come from. As we saw in Chapter 1, DNA ancestry testing can quite accurately indicate our continental ancestry, but this is done under particular assumptions that overlook the fact that non-Africans also have an ultimate ancestry in Africa. Under such assumptions we can specify local ancestries for individuals, but we should not forget that we are a recent species and that we all descend from the same ancestral, African populations. So if you want to trace your "deep roots," there is no need to take a DNA ancestry test: these are found somewhere in Africa. Ultimately, we are all Africans and we are all related through common ancestors who originated there. In this sense, human ancestry is collective and global. Taking a DNA test to find an ancestry in more recent times anywhere else in the world—in other words, an individual local ancestry—is rather arbitrary. We all have ancestors from every time and space point in our species' evolutionary history. There is no objective way to prioritize one of these points. It is like claiming that you have ancestry in the place where your grandparents were born, ignoring your great-grandparents and anyone else before them. You can do this, but there is no way to argue that this is not an arbitrary choice.

7

More Related Than Distinct

Human DNA Diversity

Several studies of human genetic diversity—beginning with a landmark 1972 study by evolutionary geneticist Richard Lewontin—have consistently shown that the vast amount of human genetic variation is found within regional groups rather than among them (Table 7.1).[1] The general conclusion from these studies is that if we take any two individuals from any two different regions, such as two continents, they will not be more different from each other than each one of them will be from an individual from the same region. In other words, most of the genetic differences are found at the individual level and are not generalizable at the population level. It is only 10%, or less, of the variation that might allow the distinction of any two individuals as belonging to two different continental regions. Therefore, it is not possible to distinguish between any two individuals as members of different continental human groups, based on their genetic differences, because these differences are so small that they are insufficient to differentiate between them. Figure 7.1 represents in a very simple form the misconception about human genetic diversity between continental groups (left) and a more accurate representation of this diversity (right). The representation on the right shows that most genetic variation among continental groups is overlapping, and not distinctive, as in the representation on the left.

There is an immediate and clear consequence of this: the average differences in DNA variation between any two populations do not entail much for

[1] Lewontin, R. C. (1972) The apportionment of human diversity. *Evolutionary Biology*, 6, 381–398; Barbujani, G., Magagni, A., Minch, E., and Cavalli-Sforza, L. L. (1997). An apportionment of human DNA diversity. *Proceedings of the National Academy of Sciences*, 94(9), 4516–4519; Jorde, L. B., Watkins, W. S., Bamshad, M. J., Dixon, M. E., Ricker, C. E., Seielstad, M. T., and Batzer, M. A. (2000). The distribution of human genetic diversity: A comparison of mitochondrial, autosomal, and Y-chromosome data. *The American Journal of Human Genetics*, 66(3), 979–988; Rosenberg, N. A., Pritchard, J. K., Weber, J. L., Cann, H. M., Kidd, K. K., Zhivotovsky, L. A., and Feldman, M. W. (2002). Genetic structure of human populations. *Science*, 298(5602), 2381–2385; see also Feldman, M., and Lewontin, R. C. (2008). Race, ancestry, and medicine. In B. Koenig, S. Lee, and S. Richardson (Eds.), *Revisiting race in a genomic age* (pp. 89–101). New Brunswick, NJ: Rutgers University Press.

Ancestry Reimagined. Kostas Kampourakis, Oxford University Press. © Oxford University Press 2023.
DOI: 10.1093/oso/9780197656341.003.0007

Table 7.1 Studies of Variation within and between Human Groups

Study	No. of Variants	No. of Regions	Variation within Groups within Regions (Vwg)	Variation among Groups within Regions (Vag)	Variation within Regions (Vwg+Vag)	Variation among Regions (Var)
Lewontin 1972	17	7	85.4%	8.3%	93.7%	6.3%
Barbujani et al. 1997	109	4–5	84.4%	4.7%	89.1%	10.8%
Jorde et al. 2000	60	3	87.9%	1.7%	89.6%	10.4%
Rosenberg et al. 2002	377	7	94.1%	2.4%	96.5%	3.6%

the differences between any two individuals from these populations. If the differences in DNA variation in populations were significant, as it is the case in the left part of Figure 7.1, then we could make inferences about the features that the respective individuals could have, and how much they might differ from one another. But as the overall variation is overlapping, as shown in the right part of Figure 7.1, any such inferences are invalid. Two individuals from

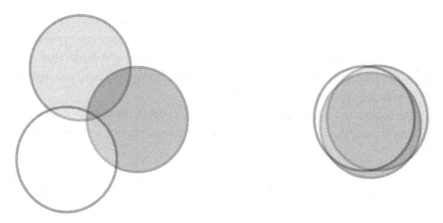

Figure 7.1 Left: The apportionment of human genetic diversity, where most variation is found between continental regions. This view is mistaken. Right: A more accurate view of the apportionment of human genetic diversity, where most variation is found within continental regions.

MORE RELATED THAN DISTINCT   125

any two populations might differ in some features and be similar in many others. If we want to compare them, we would have to study their own actual DNA variation, not make inferences about that on the basis of which population they presumably belong to. But then how is it possible for DNA ancestry tests to assign individuals to continental groups, which might be perceived to correspond to races, with high accuracy? As I explain in subsequent sections, this is a probabilistic assignment that is possible to make because of particular patterns of DNA variation in particular populations—which is a fact about populations—and not because there is much DNA variation that is distinctive of the members of a continental group, or race for some—which is a fact about individuals.[2] Simply put, we should not make inferences about individuals from facts about populations.

It is important to keep in mind that geneticists who study human DNA variation almost exclusively study populations. The models and other complex mathematical tools they use were developed to measure and analyze variation at the population level, such as measuring the frequencies of particular DNA variants within them. In contrast to genetic ancestry companies, there is no focus on individuals for population geneticists. Of course, population data emerge from the accumulation of individual data. However, what is studied and analyzed are the populations, not the individuals from which they consist. The respective studies aim at testing hypotheses, and their results are characterized by specific measures of uncertainty, and values that indicate the strength of statistical support, given the sample sizes. As geneticist Mark Jobling and his colleagues have explained, "When we recruit participants for our studies, our aim is to publish papers based on population data, but we also promise individual-based information in return for participation. We struggle to balance the need for rigour in the information we provide to an individual about their ancestry against the possibility that it is too disappointingly vague or equivocal to be of any interest at all." Jobling and colleagues thus pointed out a conflict that exists between population genetics and individual-based genetics, arguing that "although the tests themselves are reliable, the interpretations are unreliable and strongly influenced by cultural and other social forces."[3]

---

[2] Appiah, K. A. (2018). *The lies that bind: Rethinking identity*. London: Profile Books, p. 120.
[3] Jobling, M. A., Rasteiro, R., and Wetton, J. H. (2016). In the blood: The myth and reality of genetic markers of identity. *Ethnic and Racial Studies*, 39(2), 142–161.

126 ANCESTRY REIMAGINED

Does this mean then that there is no value in using continental regions, or races, as proxies for medical purposes? Yes, indeed! Even though it is true that some diseases are found more often in particular populations and geographical areas rather than others, this is independent of race. For instance, sickle-cell anemia is not only found in high frequency among people of African ancestry, as commonly argued; it is also found in high frequency in Greece, Turkey, the Middle East, and India. What explains the high frequency of sickle-cell anemia is not race, but the corresponding high incidence of malaria in those areas. People who are heterozygotes for the alleles that are related to sickle-cell anemia seem to be more protected than others from malaria caused by the parasites *Plasmodium falciparum*.[4] This, in turn, results in the maintenance of the respective allele in relatively high frequency in these populations, because heterozygotes (that is, people having a sickle cell and a "normal" allele) have an advantage as they suffer neither from sickle-cell anemia nor from malaria.

Geneticists Giorgio Sirugo, Sarah A. Tishkoff, and Scott M. Williams recently explored how the concept of race has been used and misused in medicine, and they concluded: "We have argued that although 'race' and genetic ancestry both can play a role in disease disparity, they are often independent of each other. 'Race' may be seen as a proxy of some genetic differences (but we argue it is a poor one, as clearly shown in admixed populations), but in other situations it may be a risk factor independent of the genetic variation that exists because it is a proxy for racial discrimination and social inequities."[5] A similar view was recently expressed by 16 medical professionals in a leading medical journal. Their view is that "Although race/ethnicity correlates with genetic ancestry, it captures different information. Race and ethnicity are self-ascribed or socially ascribed identities and are often 'assigned' by police, hospital staff, or others on the basis of physical characteristics. Genetic ancestry is the genetic origin of one's population. Although race/ethnicity may capture information about the likely presence of certain genetic variants, ancestry is a better predictor." However, they believed that "it is inappropriate to simply abandon the use of race and ethnicity in biomedical research and

---

[4] This may be due to an arrest of growth of the parasites found within sickle- red-blood cells maintained at low oxygen concentrations because of the synthesis of an abnormal hemoglobin (HbS). See Archer, N. M., Petersen, N., Clark, M. A., Buckee, C. O., Childs, L. M., and Duraisingh, M. T. (2018). Resistance to Plasmodium falciparum in sickle cell trait erythrocytes is driven by oxygen-dependent growth inhibition. *Proceedings of the National Academy of Sciences, 115*(28), 7350–7355.

[5] Sirugo, G., Tishkoff, S. A., and Williams, S. M. (2021). The quagmire of race, genetic ancestry, and health disparities. *The Journal of Clinical Investigation, 131*(11), e150255.

clinical practice, since these variables capture important epidemiologic information, including social determinants of health such as racism and discrimination, socioeconomic position, and environmental exposures."[6]

In short, if race matters, it is not due to the genetic differences but due to the related racial discrimination and inequalities. The differences among races that really matter are at the level of society, not at the level of DNA. That we rely on some differences in visible biological features that can be used to probabilistically assign people to social groups, because they are found in some groups more often than in others, does not mean that there are essential biological differences among these groups, and that these groups are "natural." However, there are people who still argue that races exist, or at least that distinct human populations exist even if these do not correspond to races (see the last section of the present chapter).

Let us now look in some detail how scientists have actually studied human DNA variation in populations and what inferences they can make from it.

Does DNA Variation Reflect the History and Geography of Human Populations?

There have been several studies of the geographical distribution of human DNA variation, all of which have yielded useful information both about the present and about the past. What is common in all of them is that they study the genetic variation of populations as we briefly discussed in Chapter 6: we can collect DNA samples, analyze the data, and compare the frequencies of particular variants in particular populations. Generally speaking, there exist two kinds of methods for doing such an analysis. These methods differ in the mathematics they rely upon, and their detailed consideration falls outside the scope of the present book.[7] However, it is interesting to consider them briefly in order to understand their potential and their limitations.

In the first kind of methods, populations are *predefined*: the researchers decide in advance which populations they want to compare to one another.

---

[6] Borrell, L. N., Elhawary, J. R., Fuentes-Afflick, E., et al. (2021). Race and genetic ancestry in Medicine: A time for reckoning with racism. *The New England Journal of Medicine, 384*(5), 474–480.

[7] See Pritchard, J. K., Stephens, M., and Donnelly, P. (2000). Inference of population structure using multilocus genotype data. *Genetics, 155*(2), 945–959. All these methods are based on particular mathematical models and assumptions, on which the findings of population genetics studies are strongly dependent. Therefore, they should be carefully considered, and any inferences about history and ancestry should be made with caution.

128 ANCESTRY REIMAGINED

One way to do this is to consider continental groups: for instance, Africans, Americans, Asians, and Europeans. It is also possible to arrive at a finer level of detail, if individuals are assigned to subcontinental populations, on the basis of their nationality, ethnicity, language or other cultural characteristics. In this case, the assumptions that are made in the beginning about how to individuate populations and distinguish them from one another have important implications for the results. An a priori assignment of individuals to populations can be very subjective, and different assignments can yield different results. Let us see why.

Population geneticist Luigi Luca Cavalli-Sforza was a pioneer of studies of human genetic variation and believed that it was possible to rely on it for understanding historical events. In 1994, together with colleagues Paolo Menozzi and Alberto Piazza, he published a major synthesis of the then available data on human genetic diversity. In that book, among other questions, they addressed the question of whether the spread of agriculture in Europe, between 9000 and 6000 BCE, was due to the expansion of farmers themselves (what they called a demic expansion), or due to the diffusion of agriculture as a technology (what they called a cultural diffusion). They argued that different kinds of observations, such as archaeological and ethnographic ones, supported the demic expansion model, according to which farmers expanded to Europe from the Middle East. But they emphasized that what also agreed with the demic expansion model and partial interbreeding with local hunter-gatherers was the then current geographic distribution of genes in Europe that formed a cline.[8] A cline is a geographic gradient, in which there is a gradual change in the frequencies of single nucleotide polymorphisms (SNPs) (or alleles) across geographical space. In such a case, the frequency is high in one region and gradually declines the further away one moves from that region. Clines of genetic variation can be thought of as being analogous to temperature lines on a weather map, which connect places that have the same temperature. Similarly, on a genetic cline there are lines that connect places that have the same SNP (or allele) frequencies.

The method Cavalli-Sforza and his colleagues used for finding such clines was principal component analysis (PCA). PCA is a method that allows multidimensional information to be displayed graphically. The central idea is essentially the one illustrated in Figure 6.1. PCA allows the reduction

[8] Cavalli-Sforza, L. L., Menozzi, P., and Piazza, A. (1994). *The history and geography of human genes*. Princeton, NJ: Princeton University Press, p. 108; see also his book on the topic intended for a broader audience. Cavalli-Sforza, L. L. (2001). *Genes, peoples and languages*. London: Penguin Books.

of multiple dimensions to two or three dimensions that we can represent graphically and understand, with minimum loss of information. Thus, it is possible to construct maps that summarize information of several different DNA variants with similar geographical distribution. When the individuals sampled are located across a continuum in space and the ones that are closer to one another geographically are also more genetically similar, then the outcome will be a cline. However, as geneticist John Novembre and his colleagues have noted, these patterns should be interpreted with caution. In order to test whether the population expansion that Cavalli-Sforza and colleagues had inferred from genetic data was indeed the case, Novembre and colleagues performed computer simulations. Their conclusion was that such highly structured patterns can be mathematical artifacts, and thus not real. They also emphasized that even if the patterns of genetic variation in Europe identified by Cavalli-Sforza and colleagues were correct, the Neolithic expansion is just one of several potential explanations.[9]

In a subsequent study, Novembre and colleagues studied data simulated under various models related to Neolithic farmer expansion with various levels of interbreeding between them and resident hunter-gatherer populations. They found that when the rate of interbreeding between the incomers and the residents was low, there was a gradient in a direction perpendicular to the axis of the Neolithic expansion, rather than alongside it. They concluded their article by noting that "Our study emphasizes that PC . . . should be viewed as tools for exploring the data but that the reverse process of interpreting PC . . . maps in terms of past routes of migration remains a complicated exercise. Additional analyses—with more explicit demographic models—are more than ever essential to discriminate between multiple explanations available for the patterns observed in PC . . . maps." Simply put: PCA is a good tool for exploratory data analysis, but this does not mean that the interpretation of the results is simple and straightforward.[10] Please keep in mind how wisely the researchers advise caution in the interpretation of the results.

Novembre was also a lead author in another study that aimed to explore whether populations within Europe were distinct at the DNA level or not, as

---

[9] Novembre, J., and Stephens, M. (2008). Interpreting principal component analyses of spatial population genetic variation. *Nature Genetics, 40*(5), 646–649.

[10] François, O., Currat, M., Ray, N., Han, E., Excoffier, L., and Novembre, J. (2010). Principal component analysis under population genetic models of range expansion and admixture. *Molecular Biology and Evolution, 27*(6), 1257–1268.

130 ANCESTRY REIMAGINED

well as how precisely individuals could be assigned to particular geographic locations on the basis of DNA data alone. To achieve this, the researchers studied 500,568 SNPs from a sample of 3,192 Europeans. As the samples had been collected in two places (London, UK, and Lausanne, Switzerland), the researchers used the country of origin of each individual's grandparents to determine the geographic location that best represented their ancestry. If this information was not available, they used the self-reported country of birth for this purpose. This resulted in a sample of 1,387 individuals (I consider the sampling process of this study in the next chapter). Based on these data, they used PCA to produce a two-dimensional visual summary of the genetic variation, which resembled a geographic map of Europe. In this analysis, no distinct clusters emerged, which entails that the populations considered existed along a continuum. However, individuals from the same geographic region clustered together and even some recognizable geographical features such as the Iberian and the Italian peninsulas, were distinguishable on the PCA plot. The researchers noted that it was even possible to distinguish among the French-, the German- and the Italian-speaking groups within Switzerland, as well as between Ireland and the United Kingdom. They also argued that European DNA can be very informative about geographical origins, as it was possible to place individuals who reported grandparental origins from a particular area within a certain distance of a few hundred thousand kilometers from that area based on DNA data alone. [11]

In short, it is argued that PCA can help researchers distinguish among individuals from predefined populations with high geographical resolution, if the sampling is dense and the numbers of DNA markers studied is large. Let us now see what scientists can find if populations emerge from data, and by consequence whether clusters of human DNA variation exist.

Is Human DNA Variation Clustered or Continuous?

When populations are not predefined, the aim is to let clusters (which correspond to human groups) themselves emerge from the data, without any preconceived notions. A key concept in this approach is population structure. This concept refers to the pattern by which genetic variation is distributed

[11] Novembre, J., Johnson, T., Bryc, K., Kutalik, Z., Boyko, A. R., Auton, A., et al. (2008). Genes mirror geography within Europe. *Nature*, *456*(7218), 98–101.

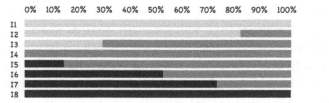

Figure 7.2 How individuals are represented in the results of STRUCTURE. Each horizontal line corresponds to one individual (I1–I8). The color of each line corresponds to the group (X, Y, or Z) to which each individual is more likely to belong. Individuals I1, I4, and I8 have 100% probability to belong to groups X, Y, and Z, respectively. However, the other individuals are more or less likely to belong to two groups. For instance, individual I3 has a 30% probability to belong to group X and a 70% probability to belong to group Y.

across a geographic area. Such patterns emerge because particular DNA variants are found in particular regions due to past processes such as migration and selection. The software used to analyze DNA data produces various possible clusters of individuals, which can then be compared to other kinds of groupings based on different kinds of data (e.g., linguistic, archaeological) available for the respective populations. One of the most popular methods used for identifying patterns of genetic variation is called STRUCTURE, developed by geneticist Jonathan Pritchard. This method draws on DNA data to identify groups of individuals that have distinctive allele frequencies. Populations are not defined before the analysis, and the researchers can use the model for a different number of populations (K) each time, which yields different results about the (probabilistic) assignment of individuals to one or more groups. In the final graph, each individual is represented by a line, which can be partitioned into differently colored segments that represent the individual's estimated membership to the various clusters/groups (Figure 7.2). The analysis can be performed several times for the same value of K to ensure consistency, and the researchers can select the K value with the highest probability. I must note that the developers of this method warned repeatedly in their article that the estimated probabilities "should be regarded as rough guides to which models are consistent with the data, rather than accurate estimates."[12]

---

[12] Pritchard, J. K., Stephens, M., and Donnelly, P. (2000). Inference of population structure using multilocus genotype data. *Genetics*, *155*(2), 945–59.

132 ANCESTRY REIMAGINED

This method was applied in what is now a widely discussed study of human populations at the global level, with geneticist Noah A. Rosenberg and Pritchard as the lead authors.[13] Instead of using any prior information about where these individuals came from, they run STRUCTURE, giving K values from 2 to 6. When K was set at 2, one cluster (roughly) included Africa, Europe, the Middle East, and Central/South Asia, whereas the other (again roughly) included East Asia, Oceania, and America. When K was set to 3, the former cluster was broken down to two others, one corresponding to Africa and the other comprising all the other regions. When K was set to 4, the second of the two initial clusters was broken down to two others, one corresponding to America and the other (roughly) comprising all the other regions. Finally, when the K was set to 5, the outcome (again, roughly) corresponded to the five major continents: Africa, Europe, Asia (consisting of Middle East and Central/South Asia and East Asia), America, and Oceania. When K was set to 6, an additional cluster emerged that consisted largely of individuals of the Kalash group, who live in northwest Pakistan. It should be noted, as the researchers themselves mentioned, that in several populations, individuals had partial membership in multiple clusters. The fact that the genetic clusters that emerged in this study "often corresponded closely to predefined regional or population groups or to collections of geographically and linguistically similar populations," as the researchers noted, has often been considered to imply that these predefined populations exist naturally (see also Chapter 8).[14]

Fascinating as this might seem, there are important issues to keep in mind, as anthropologist Deborah Bolnick has argued. The first, as already mentioned above, is that the estimated probabilities are just rough estimates. There is no certainty that these estimates are entirely accurate; this is just a method that serves as a guide. Second, it is possible that different runs of the program may yield different results, either because there was not enough time for the program to determine the optimal clustering of the individuals, or because there were several ways for dividing individuals into a particular number of clusters that were equally probable. Finally, the program is not appropriate for all sets of data, for instance for relatively isolated populations that do not interbreed extensively. Furthermore, it should also be mentioned

---

[13] They analyzed genotypes at 377 autosomal microsatellite loci in 1,056 individuals from 52 populations of the HGDP dataset.

[14] Rosenberg, N. A., Pritchard, J. K., Weber, J. L., Cann, H. M., Kidd, K. K., Zhivotovsky, L. A., and Feldman, M. W. (2002). Genetic structure of human populations. *Science, 298*(5602), 2381–2385.

MORE RELATED THAN DISTINCT   133

that Rosenberg and colleagues presented in their published article the most probable results for each value of K, but they did not state that one of the values of K, from 1 to 6, was the most probable one.[15] In short, STRUCTURE and other clustering methods can show how population structure could be, not how it actually is.

Other researchers raised concerns about the possibility of clustering humans into groups. Geneticist Svante Pääbo and his colleague David Serre considered the criteria for the collection of samples as a potential factor that affects the results. They noted that the approach most often used to collect DNA from individuals belonging to particular ethnic groups (such as "Norwegians" or "Ashkenazi Jews") excludes from the samples studied other people that are considered to have "mixed" ancestry or belonging to populations that are considered to be "admixed." In this way, what should have been a representative sampling, becomes selective because from each geographical region only people considered to exclusively belong to particular ethnic groups are included in the study, whereas others are not (this was already mentioned in the previous section for the Novembre and colleagues' study on Europe, and I get back to it in Chapter 8). Pääbo and Serre also pointed to an important discrepancy between previous studies. On the one hand, studies of human genetic diversity in particular areas with large numbers of individuals covering a substantial part of these areas had generally revealed clines, that is, gradients of DNA variant frequencies. On the other hand, studies of human genetic variation at the global level have often found clusters that roughly correspond to continents, thus suggesting a genetic diversity that corresponds to racial classification. According to Pääbo and Serre, there were two possible explanations for this discrepancy: either genetic diversity is clinal within continents and discontinuous between them; or these differences are due to how samples are selectively collected based on ethnic, or otherwise culturally defined populations, rather than being indiscriminately collected solely based on geography. In order to answer these questions, Pääbo and Serre first created and compared two datasets of equal size: one with individuals from sub-Saharan Africa, Southeast Asia, and Western Europe, and a second with individuals more widely distributed across the globe, such as Africa, Asia, Europe, Oceania, and Native

---

[15] Bolnick, D. A. (2008). Individual ancestry inference and the reification of race as a biological phenomenon. In B. Koenig and S. Soo-Jin Lee (Eds.), *Revisiting race in a genomic age*. New Brunswick, NJ: Rutgers University Press, pp. 70–85.

134 ANCESTRY REIMAGINED

Americans.[16] What they found for the first sample were two discrete continental units of diversity, one consisting of Africans and the other of non-Africans. This was not found in the other sample that they considered to better represent the geographic distribution of humans across continents.

Then Pääbo and Serre turned to the HGDP data used in the Rosenberg and colleagues' study. In that study, a model had been used in which the allele frequencies in the inferred populations at each locus were correlated with each other. In the new analysis, of a reduced number of individuals,[17] another model was used in which allele frequencies in the inferred populations were considered to be independent of each other. The results of Pääbo and Serre indicated four major clusters of human variation, similar to the Rosenberg and colleagues' study. However, there were also important differences. First, most individuals were found to have multiple ancestries. Second, and contrary to the finding of distinct continental clusters, three major geographical gradients were found: a north–south gradient in Africa; in Eurasia a gradient with western European individuals at one end and Southeast Asian individuals at the other; and in America there was also a geographical gradient with all individuals being admixed along a continuum between the Asian/Oceanian end of the Eurasian gradient and what was only found in Native Americans. The main conclusion was that the discrete clusters might be caused by discontinuities in sampling, and that therefore, it is better to sample individuals from as many places as possible when the aim is to study human genetic diversity at the global level.[18]

In a subsequent study, Rosenberg, Pritchard, and their colleagues tried to respond to these criticisms and confirm their previous findings. They first pointed out that the analysis of Serre and Pääbo was based on too little data for making the clustering evident. They also noted that as their own previous study and the Serre and Pääbo study differed in the geographic dispersion of samples, as well as the sample size and in the assumptions about allele frequency correlations, it was not possible to attribute the different results exclusively to one of these variables. What Rosenberg and colleagues concluded in

[16] One dataset was based on 89 individuals from 15 populations from data used in previous studies (30 individuals from sub-Saharan Africa, 29 individuals from Southeast Asia, and 30 individuals from Western Europe); the other dataset was based their analysis of DNA samples from 90 individuals (20 individuals from Africa, 36 individuals from Asia, 16 individuals from Europe, 13 individuals from Oceania, and 5 individuals from Native American groups).

[17] Pääbo and Serre created three subsamples of 261 individuals from the 1,066 individual data set, such that five individuals were sampled from each of the 52 populations in the HGDP panel.

[18] Serre, D., and Pääbo, S. (2004). Evidence for gradients of human genetic diversity within and among continents. *Genome Research*, 14(9), 1679–1685.

MORE RELATED THAN DISTINCT 135

their new study was that the level of geographic dispersion of the sample was the only factor that seemed to have a relatively small effect on the clustering results. Therefore, they concluded that they found no reason to interpret the clusters inferred in their 2002 study as artifacts of sampling. Rosenberg and colleagues also concluded that whereas allele frequency differences generally increase gradually with geographic distance, there exist small discontinuities across geographic barriers that allow clusters to be produced. In other words, human genetic diversity consists of both clines and clusters. While insisting on the existence of clusters, they also noted: "Our evidence for clustering should not be taken as evidence of our support of any particular concept of 'biological race.'"[19]

Another study, by Li and colleagues, aimed to analyze SNP data from the HGDP panel, without considering the population identity of participants.[20] They did this using a new software, *frappe*, similar to STRUCTURE. For $K = 5$, there was a separation into five continental groups, similar to the Rosenberg and colleagues 2002 study. For $K = 6$, there was a separation of South/Central Asia from the Middle East and Europe. For $K = 7$, there was a separation of Middle Eastern populations from European populations. Using PCA, they also found that the first and second principal components separated the populations into the usual continental groups. The first principal component primarily represented the contrast between sub-Saharan Africans and non-Africans, whereas the second principal component represented the East-West difference in Eurasia. The researchers noted that their analyses confirmed that most genetic variation is found within human populations. However, they also pointed out that between-population variation is such that it is possible to figure out the continental ancestry of individuals from their DNA.[21]

Therefore, we see that to a certain extent it is possible to arrive at the same conclusions independently, for instance regarding the clustering of human populations in the usual continental groups. However, we also see that different initial assumptions and different uses of the same data can lead to

---

[19] Rosenberg, N. A., Mahajan, S., Ramachandran, S., Zhao, C., Pritchard, J. K., and Feldman, M. W. (2005). Clines, clusters, and the effect of study design on the inference of human population structure. *PLoS Genetics*, *1*(6), e70.

[20] They studied 938 unrelated individuals from 51 populations (they considered northern and southern Han Chinese as a single population) for 642,690 common SNPs.

[21] Li, J. Z., Absher, D. M., Tang, H., Southwick, A. M., Casto, A. M., Ramachandran, S., et al. (2008). Worldwide human relationships inferred from genome-wide patterns of variation. *Science*, *319*(5866), 1100–1104.

136 ANCESTRY REIMAGINED

different results. This does not entail that we should question the science underlying these analyses. Quite the contrary, it shows that scientists are aware of how different factors can affect the results, and so they explicitly consider their impact. This is what Rosenberg and colleagues did in their 2005 study, eventually confirming their 2002 results. It should be clear that this is how science is done: with careful control of variables, independent testing, replication of results, and more. However, even if the groups in the aforementioned studies were allowed to emerge from the analysis, the findings are still method-driven rather than data-driven. Different methods make different assumptions and can arrive at different results. Therefore, the data do not speak for themselves, but they can be interpreted in different ways, arriving at different conclusions. As Novembre has pointed out, "STRUCTURE has become, in some sense, a victim of its own success. It is applied by default in most studies without consideration of whether the underlying model is relevant."[22]

There are two important points to keep in mind. STRUCTURE does not reveal clusters that already exist; rather, these are the outcome of a complex statistical procedure and thus of the program itself. Therefore, the program alone cannot point to the correct number of clusters, and this is why the researchers suggest that a population's history should also be taken into account. However, it is also important not to let this history, or our understanding of it, guide the conclusions—for instance, take the value $K = 5$ as more appropriate because the outcome fits well with the known continental groups. Most importantly, the nice visual images that the program produces should be considered with caution. To give one example, mathematician Daniel John Lawson and colleagues used the program ADMIXTURE, which is similar to STRUCTURE, and found that the plots of three different fictional scenarios were almost identical! Therefore, any of these three scenarios could explain the results, even though only one of them was true. Lawson and colleagues concluded that "these analyses should always be followed up with tests of specific hypotheses, using other approaches. Running STRUCTURE or ADMIXTURE is the beginning of a detailed demographic and historical analysis, not the end."[23]

---

[22] Novembre, J. (2016). Pritchard, Stephens, and Donnelly on population structure. *Genetics*, 204(2), 391–393.
[23] Lawson, D. J., Van Dorp, L., and Falush, D. (2018). A tutorial on how not to over-interpret STRUCTURE and ADMIXTURE bar plots. *Nature Communications*, 9(1), 1–11.

In short, these methods are useful in an exploratory rather than in a confirmatory sense. However, this research can be, and has been, misinterpreted. Perhaps not surprisingly, some of the research presented so far has been used to support the claim that biological races exist.

Biological Races: The Persistence of a Scientifically Flawed Idea

Journalist Nicholas Wade, in his book *Troublesome Inheritance*, and political scientist Charles Murray, in his book *Human Diversity*, have argued that the existence of distinct clusters of humans, as revealed by recent genomic science, stand as confirmation for the existence of biologically distinct races. In addition, they have argued that scientists overlook the scientific evidence about this, or refrain from talking about it, for political or other nonscientific reasons. There have been several criticisms of their views.[24] Here are I only consider two of their arguments that might seem convincing:

1. Human populations are distinct at the DNA level in ways that correspond to self-identified race and ethnicity.
2. There has been extensive natural selection since humans left Africa, which has led to local adaptation in different continents and which has thus further diversified human populations to form distinct races.

Regarding the first point, Wade and Murray cited the studies discussed in the previous section to suggest that contemporary human populations are genetically distinct in ways that match continental divisions, and therefore races. Already on page 2 of his book, Wade wrote that "the populations on each continent have evolved largely independently of one another as each adapted to each regional environment. Under these various local pressures, there developed the major races of humankind, those of Africans, East Asians and Europeans, as well as many smaller groups." He also wrote that "there is usually continuity between neighboring races because of gene exchange between

---

[24] See, for instance, Fuentes, A. (2014). A troublesome inheritance: Nicholas Wade's botched interpretation of human genetics, history, and evolution. *Human Biology, 86*(3), 215–219; Marks, J. (2014). Review of a Troublesome inheritance by Nicholas Wade. *Human Biology, 86*(3), 221–226. A book-length critique is DeSalle, R., and Tattersall, I. (2018). *Troublesome science: The misuse of genetics and genomics in understanding race*. New York: Columbia University Press; Ball, P. (2020). The gene delusion. *New Statesman*, June issue, "A World in Revolt.".

138 ANCESTRY REIMAGINED

them. Because there is no clear dividing line, there are no distinct races—that is the nature of variation within a species. Nonetheless, useful distinctions can be made." He thus noted that Africans, East Asians, and Europeans are "the three racial groups that everyone can identify at a glance," also adding Australian aborigines and American Indians to complete the set of five races that correspond to the five main continents. His justification for the choice for this number was this: "to keep things simple, the five-race, continent-based scheme seems the most practical for most purposes."[25] A justification can hardly seem more arbitrary than this, but Wade also cited the Rosenberg and colleagues' studies, as well as the Li and colleagues' study, all discussed in the previous section, as providing evidence for the clustering of humans at the continental level.

Murray developed similar arguments. He described as the scientific orthodoxy the view that genetic differences among populations are insignificant and that humans left Africa too recently for important differences to evolve. At the end of his introduction in Part II of his book, he wrote that "Part II sets out to convince you that the orthodoxy about race is scientifically obsolete." But he also noted that "The orthodoxy is not wrong altogether but goes too far when it concludes that race is biologically meaningless." He agreed that the term "race" should be replaced by terms such as ancestral population; but he nevertheless argued that race and ethnicity are not just cultural constructs; otherwise it would not be possible to infer the ancestral populations (or races, or ethnicities) to which individuals are related simply by studying their DNA. He concluded his review of the Rosenberg and colleagues, and Li and colleagues, studies by writing that "The material here does not support the existence of the classically defined races, nor does it deny the many ways in which race is a social construct. . . . Genetic differentiation among populations is an inherent part of the process of peopling the Earth. It is what happens when populations successively split off from parental populations and are subsequently (mostly) separated geographically."[26] In short, both Wade and Murray argued that the studies reviewed in the previous section stand as confirmation for point 1 above.

Let us now consider their arguments for point 2: that the races thus identified evolved independently due to dealing with different environments

---

[25] Wade, N. (2014). *A troublesome inheritance: Genes, race and human Hhstory*. New York: Penguin Press, pp. 2, 92–93, 100.

[26] Murray, C. (2020). *Human diversity: The biology of gender, race, and class*. New York: Twelve, pp. 132, 135, 156–157.

that resulted in natural selection toward different evolutionary outcomes. To make this point, Wade considered in detail two studies. The first one, led by Pritchard, provided evidence in the human genome about recent positive selection. Wade wrote that "In each race Pritchard found about 200 genetic regions that showed a characteristic signature of having been under selection (206 in Africans, 185 in East Asians, and 188 in Europeans). But in each race, a largely different set of genes was under selection, with only minor overlaps" (there were 38 regions common between Africans and East Asians, 35 between Europeans and Africans, 55 between East Asians and Europeans, and 10 common among all three groups). This was, for Wade, the evidence for the independent evolution, and therefore the distinctiveness of races. To make his case, he also included in his book a redrawn version of a figure from the article by Pritchard and colleagues that Wade cited. Wade wrote in the caption of the figure "Regions of the genome that are highly selected in the three major races."[27] As he wrote, the regions that were under selection in each race were "largely different" and the overlaps among them were only minor. Indeed, given the numbers mentioned above, it seems that more genomic regions under selection are distinct than shared among races. But there were also 26,374 regions (on the top left of the original figure, neither included in Wade's representation of it, nor discussed by him) that were not found to be under selection in any of the three groups. I think that the fact that only 2.7% of the studied genome segments were found to be under selection tells us something about the significance of Wade's argument.

There are at least two problems with Wade's argument. The first one has to do with the data used that came from the HapMap Project, which is based on a few populations (Table 6.1).[28] Whether these samples suffice to represent Wade's three main races is a big question, but one I consider in the next chapter. The second problem has to do with how Wade represented and interpreted the findings of Pritchard and colleagues. What that study intended to do was to find sites on the genome where there is strong, very recent, selection for alleles that are not "fixed" in a population (i.e., are found in 100% frequency). They also noted that "Furthermore, signals of selective sweeps in progress indicate the presence of genetic variants that must have

[27] Wade, N. (2014). *A troublesome inheritance: Genes, race and human history.* New York: Penguin Press, pp. 103–104.

[28] As Pritchard and colleagues wrote, they analyzed 800,000 polymorphic SNPs in a total of 309 unrelated individuals: 89 Japanese and Han Chinese individuals from Tokyo and Beijing, respectively, which were denoted as East Asian (ASN), 60 individuals of northern and western European origin (CEU), and 60 Yoruba (YRI) from Ibadan, Nigeria.

140 ANCESTRY REIMAGINED

Box 7.1 What Is a Selective Sweep?

A selective sweep occurs when a new DNA variant that emerges through a mutation is beneficial for its bearers, and because it has a large effect on survival and reproduction it becomes prevalent in a population. If selective sweeps in different populations are related to adaptation to local conditions, then it is expected to find large frequency differences of particular DNA variants between populations. This might seem as evidence that selection is driving the differentiation of human populations. Generally speaking, there are two kinds of selective sweeps: hard sweeps, where a single haplotype has increased in frequency due to recent selection for a beneficial variant; and soft sweeps, where there is selection for one or more variants that have existed in the population for some time. The main feature of a soft selective sweep is that multiple haplotypes can each have a favorable variant and therefore increase in frequency at the same time. Hard selective sweeps do not seem to be the dominant mode of linked selection in the human genome.[1] This is actually, and spectacularly, the case for skin color. As we saw in Chapter 3, many genes seem to be implicated in the development of this trait. Thus, the geographic variation in skin color is not exclusively driven by hard selective sweeps in a few genes. Rather several different genes are involved, and different types of selection have acted together and produced the variation in skin color that we currently observe.[2]

[1] Jobling, M., Hollox, E., Hurles, M., Kivisild, T., and Tyler-Smith, C. (2013). *Human evolutionary genetics* (2nd ed.). New York: Garland Science, pp. 203–205; Hartl, D. L. (2020). *A primer of population genetics and genomics* (4th ed.). New York: Oxford University Press, pp. 117–118.

[2] Rocha, J. (2020). The evolutionary history of human skin pigmentation. *Journal of Molecular Evolution, 88*(1), 77–87.

---

some significant effect on human phenotypic variation."[29] (See Box 7.1 for what a selective sweep is.) Pritchard and colleagues noted in their conclusion that "Though most selected regions are not shared across populations, there is still a clear excess of shared selective events. Indeed, since we have incomplete power to detect selection, it is likely that we tend to underestimate the

---

[29] Voight, B. F., Kudaravalli, S., Wen, X., and Pritchard, J. K. (2006). A map of recent positive selection in the human genome. *PLoS Biology, 4*(3), e72.

MORE RELATED THAN DISTINCT   141

degree of sharing across populations."[30] This simply means that the findings are not as definitive as Wade presented them to be, and that the shared DNA sequences were more than what they estimated. Not surprisingly, Murray considered that study, too. Whereas the researchers wrote, as already mentioned, that the degree of sharing might have been underestimated, Murray wrote in his book exactly the contrary: "In the authors' judgement, the degree to which selection occurred independently is probably underestimated by these percentages," quoting a different passage from the article than the one I quoted above.

Wade also looked at another study led by geneticist Pardis Sabeti. The study identified 412 regions in the genome under selection: 140 in Europeans only, 140 in East Asians only, and 132 in Africans only.[31] These are data from a table (S2) in the supplementary materials of the article that presents a "list of novel and previously discovered regions under positive selection." Wade explained the absence of any overlap, that is, finding regions that are being selected in two or more populations, as was found in the study by Pritchard and colleagues, as being due to the genome scanning methods used in that study, "which depended in part on looking for sites at which the three races differed."[32] However, this is not what Sabeti and colleagues thought. There are many tests to detect positive natural selection that look for different kinds of signals left behind when an adaptive trait rises in frequency. Some of the methods are specifically designed to look for the signal of differences between populations as a trait emerges in one population but not another. By definition, therefore, the test itself will be biased toward highlighting the differences, rather than any similarities, between population groups. Even with this consideration, however, the genome scans of Sabeti and colleagues still revealed a substantial amount of overlap in the identified traits between various population groups. This means that the signals of natural selection impact many populations, demonstrating that the global community as a whole has more in common than one test or method may immediately indicate. "The nature of many of these tests is specific and biased—i.e., they detect differences between populations—and more work must be done to fully understand both our similarities and differences as humans. Moreover,

[30] Voight, B. F., Kudaravalli, S., Wen, X., and Pritchard, J. K. (2006) A map of recent positive selection in the human genome. *PLoS Biology*, 4(3), e72.

[31] Grossman, S., Andersen, K., Shlyakhter, I., Tabrizi, S., Winnicki, S., Yen, A., et al. (2013). Identifying recent adaptations in large-scale genomic data. *Cell*, *152*(4), 703–713.

[32] Wade, N. (2014). *A troublesome inheritance: Genes, race and human history*. New York: Penguin Press, p. 105.

142   ANCESTRY REIMAGINED

while reviewing the regions that ranked high for natural selection, our test revealed that the majority of genes under natural selection include those that drive differences in skin-deep features, such as skin color and visible patterns on the ectoderm (e.g., hair, sweat), as well as immune system characteristics. The processes of evolution and natural selection, therefore, appear to be affecting, in large part, interactions with the outside environment" (Pardis Sabeti, personal communication).

Murray referred to the study by Sabeti and colleagues in his own book, but he made no reference to the lack of overlaps among the three populations studied. However, he wrote about another study that provided more evidence about recent natural selection in humans. Murray wrote that the authors applied their method "to six populations that have low levels of historical admixture: three of sub-Saharan Africans (two from West Africa, one from Kenya), one of non-Latino whites in Utah, one of Japanese in Japan, and one of Amerindians in Peru. Their method identified 1,927 'distinct selective sweeps,' of which 1,408 were ones not previously identified." Based on this and other studies, Murray concluded by writing that "the regions [in the genome] under selective pressure since the dispersal from Africa are in fact extensive" and that a "substantial amount of the genome has been influenced—both directly and indirectly—by selection." Murray concluded the chapter by noting that evolution since humans left Africa was not just skin deep.[33] As he was not explicit about this, I can only infer that it is this evidence about recent selection that made him write that the orthodoxy about race (that races are not real biologically speaking) is "scientifically obsolete." The article that Murray cited concluded that "the vast bulk of human adaptation is occurring as a consequence of soft sweeps." This was the main conclusion of the researchers in their attempt to "address the controversy over the impact of adaptation on human genomic variation by conducting a genome-wide scan for both hard and soft selective sweeps across human populations." What they found were 1,927 distinct selective sweeps, 92.2% of which were classified as soft ones. Among these, only 36.4% were specific to particular populations: 33.5% were exclusive to African populations, and 66.5% were exclusive to non-African populations. Murray neither mentioned nor commented on this result that most sweeps were found in non-African populations.[34]

---

[33] Murray, C. (2020). *Human diversity: The biology of gender, race, and class.* New York: Twelve, pp. 176–177, 181.

[34] Schrider, D. R., and Kern, A. D. (2017). Soft sweeps are the dominant mode of adaptation in the human genome. *Molecular Biology and Evolution, 34*(8), 1863–1877.

But there is another important piece of information about which Murray was not clear: that the six populations in this study came from Phase 3 of the 1KPG, which means that three of those were the populations included in the Pritchard and colleagues' study that both Murray and Wade considered: the Yoruba, the Utah, and the Japanese of the HapMap Project and eventually of the 1KGP. This is very important to note because whereas for Wade these three populations were considered as representing "races," for Murray these very same populations (minus the Han Chinese) were considered as representing "populations that have low levels of historical admixture." But is it possible for the same population sample to be considered, on the one hand, as representative of the whole continent and, on the other hand, as being relatively isolated at the genetic level? The selective use of the same data for making different points can make one wonder about the validity of these arguments.

The conclusion from all this is that whereas ancestry is about relatedness, and this is clearly the case by the studies presented in Table 7.1 (but others, too), we tend to privilege the distinctiveness that we find when we focus on the differences among human groups. But it is important not to forget that this distinctiveness stems from the study of very small portions of our DNA, in which we differ. As we humans share 99.9% of our DNA, it does not make sense to focus on clusters that we find when we analyze DNA data, as if this is all that is there. For population geneticists, it makes sense to focus on these differences; when one wants to study human DNA variation, it is differences that are interesting to figure out. But this does not entail that those differences are all that matter. Ancestry is mostly about our relatedness and we should not privilege distinctiveness and provide the grounds for racist discriminations supposedly supported by DNA data. We, humans, are a lot more related to one another than distinct.

As we saw in this chapter, sampling is a crucial issue for the studies of human DNA variation, and it is to this that we now turn.

8

Social Constructs versus "Natural Order"

Biogeographical Ancestry

In the 1970s, population geneticists realized that some genetic markers are found in higher frequency in some populations than others, and these might be used to assign individuals to particular populations. However, as we saw in Chapter 3, the use of racial and ethnic categories in order to distinguish different populations in biological and biomedical research was controversial. This led to the invention of the concept of biogeographical ancestry (BGA) as an alternative. BGA was first defined as "the component of ethnicity that is biologically determined and can be estimated using genetic markers that have distinctive allele frequencies for the populations in question (referred to as population associated alleles PAAs)." However, as philosopher Lisa Gannett has shown, even though the concept of BGA was coined in an attempt to replace the concept of (biological) race, it was nevertheless based on the very same geographical and cultural criteria that we use to distinguish among races, and so it could not escape the respective social and political connotations.[1]

BGA was based on earlier work by geneticist Mark Shriver suggesting that it might be possible to distinguish among "any two major geographically or ethnically defined populations" by selecting those DNA markers that have large frequency differences (more than 50%) among them. These markers were initially described as "population-specific alleles" (PSAs); however, they were not really specific as most of them were found in more than one population. These PSAs were used to distinguish between what were perceived as ethnically different populations within the US context: one set of markers could be used for distinguishing between African Americans and European Americans, and the other for distinguishing between European Americans

---

[1] Gannett, L. (2014). Biogeographical ancestry and race. *Studies in History and Philosophy of Biological and Biomedical Sciences, 47*, 173–184.

Ancestry Reimagined. Kostas Kampourakis, Oxford University Press. © Oxford University Press 2023.
DOI: 10.1093/oso/9780197656341.003.0008

SOCIAL CONSTRUCTS VERSUS "NATURAL ORDER"   145

and Hispanic Americans.[2] When single nucleotide polymorphisms (SNPs) were used later, the markers were given a new name: "Ancestry informative markers (AIMs) are genetic loci showing alleles with large frequency differences between populations."[3] At first thought, it might sound simple and straightforward to distinguish between different populations based on the differences in their frequencies of particular alleles (as shown in Figure 6.1). However, there are two underlying issues that are worth careful consideration.

The first one is that AIMs are not indicative of particular groups in any absolute sense (for instance, that only people in group A have allele a and only people in group B have allele b). Rather, these markers are indicative of particular groups in a probabilistic sense: they are more likely to be found in people from one group rather than in people from another group (for instance, people in group A are more likely to have allele a than people in group B, who in turn are more likely to have allele b than people in group A). So how this works is that one AIM is more likely to indicate ancestry from one region, another AIM is more likely to indicate ancestry from another region, and so on. When many AIMs are considered together, individuals can be probabilistically assigned to one of these regions, based on which AIMs they are found to have. For instance, if you have the AIMs more usually found among Europeans, it means that you are more likely to have recent common ancestors with those who are identified as Europeans, rather than with those who are identified as Africans or East Asians.

This brings us to the second important issue, which is often ignored or misunderstood. What the researchers who came up with the various sets of AIMs have done was to find which markers are found more frequently in people who live today in particular regions. I repeat and emphasize this: *in people who live today!* The various human genome diversity projects (Table 6.1) used data coming from people who live today, or have lived until recently, not from any ancestral populations. In this sense, if you are found to have particular AIMs, these indicate the probability to have a recent ancestor with people who live today in the region in which these AIMs are more frequently found, rather than with people who live today in another region in

---

[2] Shriver, M. D., Smith, M. W., Jin, L., Marcini, A., Akey, J. M., Deka, R., et al. (1997). Ethnic-affiliation estimation by use of population-specific DNA markers. *American Journal of Human Genetics, 60,* 957–964.

[3] Shriver, M. et al (2003). Skin pigmentation, biogeographical ancestry and admixture mapping. *Human Genetics, 112*(4), 387–99.

146　ANCESTRY REIMAGINED

which these markers are less frequently found. But this does not say anything about your ancestry, because one cannot know for sure that the ancestors of the people included in the sample had always lived in that region (more about this in Chapter 9). Therefore, what is described as biogeographical ancestry is not really about ancestry at all; it is rather about biogeographical relatedness, because what can be found is whether one is more or less likely to be more closely related to people in one region rather than those in another, because they have more recent common ancestors with the former than with the latter and thus share many AIMs.[4] Whenever DNA from people living today is used, the inferences made are about relatedness of contemporary people through recent common ancestry, not about ancestry per se. As I explain in Chapter 10, this is exactly the case for DNA ancestry testing: test-takers are compared to particular reference groups of people who live today.

An interesting finding with the use of AIMs is that many people are found to have ancestries from different regions, for instance 50% from Europe and 50% Africa, as was the case for Henry Louis Gates Jr. These multiple ancestries consisting of different proportions are described as ancestry proportions. Once we establish a set of markers that is characteristic of the ancestry of a particular geographically defined population (this is what being ancestry informative is about), it is possible to infer the most likely ancestry of individuals by finding the AIMs that they have. In particular, for any given genomic profile it is possible to estimate what proportion of ancestry is derived from various geographical populations. This results in estimates such as, for instance, 85% sub-Saharan African and 15% Native American, or 75% European, 15% sub-Saharan African, and 10% Native American. As I showed in Chapter 1, these are more or less the kinds of estimates that DNA ancestry companies provide.

These people who are found to have multiple ancestry proportions are described as admixed, and the phenomenon is described as *admixture*. However, there is an important underlying assumption that should be explicitly considered because it is deeply problematic. The concept of admixture assumes that there exist "pure" or "distinct" categories that can in turn become "admixed." It is not possible to have admixture if there are no distinct categories to admix. Attention! I refer to categories, not populations. Even if we accept that all human populations are admixed, we still need to rely

---

[4] For an insightful and detailed relevant discussion, see also: Gannett, L. (2014). Biogeographical ancestry and race. *Studies in History and Philosophy of Biological and Biomedical Sciences, 47*, 173–184.

SOCIAL CONSTRUCTS VERSUS "NATURAL ORDER"   147

on "pure" categories to describe their admixture. As race, gender, and law scholar Dorothy Roberts cogently put it: "We can only imagine someone to be a quarter European if we have a concept of someone who is 100 percent European."[5] Or, as anthropologist Jonathan Marks simply put it: "Looking for admixture, after all, presupposes a primordial state without it."[6] Once we do so, we imply that a person with admixed ancestry is supposed to have a mixture of two or more "pure" or "distinct" ancestry categories, which should—at least in theory—consist of people who are fundamentally similar to one another and fundamentally different from people of other ancestries. This is how essentialism re-enters the scene: admixture thus conceived is the backdoor to essentialism.

However, no ancestry category is pure or distinct in any absolute sense, as different populations do not differ in which DNA variants they carry, but rather in the relative frequencies of those variants. Therefore, it is not possible to define genetically what it means to be 100% African or 100% European. Here is then a fundamental problem with the AIMs: the criterion we use to select them has to do with how well they help us distinguish among the categories we consider a priori as pure or distinct (racial or ethnic groups). To select these AIMs, we often exclude those people who cannot be assigned to only one of those categories. Then we use the AIMs thus selected to infer the ancestry of other people. However, what guides the inferences are not the AIMs themselves, but rather the biased selection we initially did. AIMs are informative about the ancestral categories we subjectively decided that exist, and so they serve to confirm this subjective decision we initially made. As anthropologist Duana Fullwiley put it: "This correspondence comprises a biologistical construction of race in which certain raced US 'populations' ('Black'/African, 'White'/European, and 'Red'/Native American) and DNA markers with certain statistical frequencies in those populations are each posited as first principles to infer truths about the other."[7]

Perhaps the most important concern about AIMs is ascertainment: the procedure used to find those SNPs that vary among different human populations. Molecular evolutionist Robert DeSalle and paleoanthropologist Ian Tattersall have argued that AIMs are the outcome of an

---

[5] Roberts, D. (2011). *Fatal invention: How science, politics, and big business re-create race in the twenty-first century.* New York: The New Press, p. 228.

[6] Marks, J. (2012). My ancestors, myself. Aeon. https://aeon.co/essays/what-can-neanderthal-dna-tell-us-about-human-ancestry

[7] Fullwiley, D. (2008). The biologistical construction of race: Admixture technology and the new genetic medicine. *Social Studies of Science, 38*(5), 695–735.

148   ANCESTRY REIMAGINED

extreme ascertainment process.[8] This in turn can result in what is described as ascertainment bias: having a dataset that is different from the respective population it is supposed to represent, because the selection of samples resulted in some members of the population being less likely to be included than others. Perhaps then the solution could be to use whole-genome sequencing data. Well, even that might not be enough. In a recent study, the researchers used whole-genome sequencing data in order to achieve an unbiased selection of DNA variants.[9] Their conclusion was that the use of predefined sets of SNPs may arrive at an overall accurate representation of the relationships among non-African populations, but not among the African ones. The researchers also found that populations in central and southern Africa, the Americas, and Oceania each have tens to hundreds of thousands of "private" DNA variants, that is, variants that have occurred in one lineage but not in others (e.g., due to mutations). However, although some of these variants were found at relatively high frequencies, none was found at 100% frequency (when this happens, they are described as "fixed") in any continent or major geographical regions.[10] This simply means that there are no DNA variants that are indicative of a geographical region or population in any absolute sense. AIMs are not biogeographical essences and should not be presented as such.

Eventually, the most important problem with the AIMs is that instead of replacing race with ancestry, they are prone to being misinterpreted as confirming the existence of races on the basis of ancestry. AIMs are used in order to differentiate among predefined social groups, and the results depend mostly on the scientists' preconceptions about who belongs to each group. They show us that someone is, say, European or African, Greek or Nigerian, because we selected particular AIMs that we believe reflect those groups. But what the AIMs really reflect are only the choices we made about who should be considered European or African, Greek or Nigerian when we conducted the analysis to select them. The AIMs do not reveal any inherent ethnic

[8] DeSalle, R., and Tattersall, I. (2018). *Troublesome science: The misuse of genetics and genomics in understanding race*. New York: Columbia University Press, pp. 87–93; DeSalle, R., and Tattersall, I. (2022). *Understanding race*. Cambridge: Cambridge University Press.

[9] They used data from the HGDP Project, sequencing 929 genomes from 54 geographically, linguistically, and culturally diverse human populations (with an average coverage of 35x). They thus identified 67.3 million SNPs, 8.8 million indels (insertions or deletions of very small nucleotide segments, which may range from one nucleotide to tens of nucleotides), and 40,736 CNVs.

[10] Bergström, A., McCarthy, S. A., Hui, R., Almarri, M. A., Ayub, Q., Danecek, P., et al. (2020). Insights into human genetic variation and population history from 929 diverse genomes. *Science*, *367*(6484), eaay5012.

SOCIAL CONSTRUCTS VERSUS "NATURAL ORDER"   149

identity. Therefore, references to AIMs and BGA can be very misleading. As sociologist Rogers Brubaker nicely put it: "And the parental populations to which autosomal tests assign ancestral proportions may, as noted earlier, be conflated with identically labeled commonsense racial, ethnic, and national categories in such a way that the relative genetic similarity used to construct the former is take as proof of the biological existence of the latter."[11]

Let us now consider in detail why we cannot really individuate human populations.

Individuating Human Populations

Geneticist David Reich is one of the major figures in the study of ancient DNA (aDNA). His book *Who We Are and How We Got Here* provides a fascinating account of the respective studies and findings. Reich explicitly argued throughout his book against the existence of biological races. For instance, he wrote that "Today, many people assume that humans can be grouped biologically into 'primeval' groups, corresponding to our notion of 'races,' whose origins are populations that separated tens of thousands of years ago. But this long-held view about 'race' has just in the last few years been proven wrong."[12] However, he also insisted throughout the book that there exist distinct human populations, in writing that "The right way to deal with the inevitable discovery of substantial differences across populations is to realize that their existence should not affect the way we conduct ourselves. As a society we should commit to according everyone equal rights despite the differences that exist among individuals." Nobody can argue with that, but what I think is critical is how to assess differences among individuals, which Reich here seems to suggest can be inferred from the documented differences among the respective populations. A simple counterargument could be that, well, any differences at the DNA level are found in its 0.1%—why neglect the remaining 99.9%? It is not as simple as this, though.

Reich presented the evidence supporting the idea that the hybrids emerging from the interbreeding between Neanderthals and modern humans were less fertile than the offspring of modern humans. Therefore,

---

[11] Brubaker, R. (2015). *Grounds for difference*. Cambridge, MA: Harvard University Press, p. 73.
[12] Reich, D. (2018). *Who we are and how we got here: Ancient DNA and the new science of the human past*. Oxford: Oxford University Press, p. xxiv.

150   ANCESTRY REIMAGINED

it would be more likely for the offspring of modern humans rather than for the hybrids emerging from their interbreeding with the Neanderthals to survive and reproduce. In other words, there could have been natural selection against the hybrids, which in turn resulted in the low amount of Neanderthal DNA in the genomes of modern humans. After presenting the evidence that natural selection against Neanderthal DNA has resulted in the decrease of its proportion in the human genome from 3% to 6% in old times, to around 2% in later times, Reich made an astonishing point: that a study of about 30,000 African Americans has "found no evidence for natural selection against African or European ancestry." Reich went on to explain that this can be due to two possible reasons. One is that the time of divergence between Neanderthals and modern humans has been 10 times longer than between West African and Europeans, thus giving more time for biological incompatibilities to develop in the former case than in the latter. Reich then concluded that "In light of this the lack of infertility in hybrids of present-day humans may no longer seem so surprising."[13]

This may not be surprising. But what is surprising, in my view, is that Reich described the offspring of West Africans and Europeans as "hybrids of present-day humans." This can be a very misleading description, especially for nonexperts. A hybrid is defined in the Cambridge dictionary as "a plant or animal that has been produced from two different types of plant or animal."[14] Whereas, of course, West Africans and Europeans are not different species, one wonders in what sense they might qualify as different "types." How can one distinguish between the members of the same species and assign them to different groups so that their offspring can qualify as hybrids? Does this suggest that the comparison of Neanderthals to modern humans and that of modern humans to one another differs only in degree rather than kind? Modern humans and Neanderthals can be considered as different varieties of humans, or even two closely related human species, who diverged some time ago while initially interbreeding among themselves. But then, are West Africans and Europeans two equivalently diverging groups that are currently found early in their process of divergence? When I made a post about this on Twitter,[15] several geneticists responded that "hybrid" is a legitimate term to use, that any intelligent reader would understand what Reich meant, and that

---

[13] Reich, D. (2018). *Who we are and how we got here: Ancient DNA and the new science of the human past*. Oxford: Oxford University Press, pp. 49–50.

[14] https://dictionary.cambridge.org/dictionary/english/hybrid

[15] https://twitter.com/KampourakisK/status/1402159554088030209

SOCIAL CONSTRUCTS VERSUS "NATURAL ORDER"   151

what he wrote would be misleading only if taken out of context. I am happy to accept they are right. But my aim is not to criticize Reich and his book. Rather, my aim is to point out that any insistence on the existence of genetically distinct human groups should clearly explain the probabilistic nature of such an estimation, and clarify what such differences do and do not entail for the respective individuals.

This brings us to a very important issue. Geneticists sometimes seem to work under the assumption that their science is objective and value-free. In this sense, they study DNA variation, analyze it, and publish their results. These results are supposed to indicate how the world really is. As Reich put it, there are no biological races; however, distinct populations do exist. The implication of this point is that if anyone relies on these results to make the inference that if populations distinct at the DNA level exist, then biological races also exist, it is them to blame, not geneticists or their science. The problem with such a view is that it overlooks (or ignores) the inbuilt assumptions in the way the scientific analyses are conducted. These assumptions often offer support for racial discrimination, and therefore preclude the science from being value-free. One such inbuilt assumption is the notion of admixture.

Geneticists, like Reich, tell us that admixture is the rule in human evolution and the resulting DNA diversity. This is good and important in order to debunk any notions of genetic purity about social groups. However, its premises are a bit problematic. As I have already explained in the previous section, any notion of admixture, or hybrids, requires the pre-existence of "pure" or unadmixed categories. Something is a mixture only when it consists of two, or more, clearly distinct categories. Eventually, when geneticists legitimately inform us that many of us are admixed, having ancestry from several populations, they might be seen as accepting the premise that pure categories exist. Geneticists compare populations that have to be somehow identified, and so geography, race, ethnicity, or nationality is used as a proxy. But because these categories are found to be different in something within their members—differences in frequencies of DNA variants—we might end up essentializing these categories. Then, by confusing categories with populations, we might end up essentializing populations themselves. Having essentialized populations, we might then think that there have existed "pure" populations that at some point in the past became admixed. Thus, by studying admixture we could establish the existence of "pure" populations, such as races, to make comparisons. These pure, essentialized populations are now naturalized, because we distinguish among them on the basis of

152 ANCESTRY REIMAGINED

DNA variants. But we now know that there are no genetically distinct human populations; our species is very young for this. There exist geographically dispersed populations and clines of genetic variation among them. Those that are further apart on the cline will be less similar genetically than those being closer to one another. This brings us to the important issue of how we designate populations.

The designation of a population would be clear if we found sufficient criteria for accurately individuating groups from one another. This in turn requires the exclusive assignment of individuals in particular populations; we can have clearly individuated groups only if individuals can only be assigned to one group and not to others. Several different criteria can be, and have been, used for this purpose: racial (e.g., Blacks and Whites), national (e.g., French and Swiss), ethnic (e.g., Israelis and Palestinians), cultural (e.g., Protestant Irish and Catholic Irish), or geographical (e.g., African Americans and European Americans). At first sight, it might seem simple and straightforward to individuate populations based on these criteria. However, if you think hard about this, it is not necessarily so. The reason is that human populations have no strict boundaries; because members of neighboring populations have always been interbreeding, there has never been any real reproductive isolation among them.[16] Therefore, we can use any criteria we may like to distinguish between different populations, but this will always be arbitrary. As geneticist Molly Przeworski cogently put it on twitter (and I quote her because I could not have put it more clearly):

"populations" are an abstraction and not a real thing . . . Population genetics offers operational definitions of populations with which we model genetic variation: a set of individuals that draw their allele frequencies from the same set of distributions . . . But most of the time, we don't mean something all that precise. As the quip goes "An academic discipline is a set of individuals who agree not to question the same assumption." For population genetics, that assumption is a population.[17]

If populations are abstractions, it means that one can define and designate them however one wants. And indeed, this is what scientists actually do. An insightful analysis by anthropologist Amade M'Charek of a

[16] Jobling, M., Hollox, E., Hurles, M., Kivisild, T., and Tyler-Smith, C. (2013). *Human evolutionary genetics* (2nd ed.). New York: Garland Science, pp. 11–12.
[17] https://twitter.com/mollyprzew/status/1437863686786457603

SOCIAL CONSTRUCTS VERSUS "NATURAL ORDER"   153

particular forensic case has revealed that several concepts of population were at play at the same time. In particular, that was a case in which it was investigated whether a suspect of Turkish origin was guilty of the murder of another person in the Netherlands. Different populations could be conceived at the same time on the basis of different criteria, such as surnames (assuming that those having Dutch surnames were born there); genetic proximity or distance based on the country of origin (Dutch and Turkish people could be considered to form two distinct groups); racial criteria (both Dutch and Turkish people were mentioned in passing as Caucasian); national identity (people of Dutch and Turkish origin could be considered to form one local group who lived in the Netherlands, if they were all Dutch nationals); and genetic markers (depending on which markers are used for grouping people into populations, it is possible to end up with different groupings).[18] Long story short: there is no single ideal criterion for designating populations.

With this in mind, we can now understand the legitimate way in which population geneticists use the concept of admixture, without any reference in eternal, "pure" categories. The key point is that population geneticists practically define operational categories in specific contexts of interest. For example, consider the case of Neanderthal admixture. If we define "Neanderthal" to be all humans who lived in Western Eurasia 100,000 years ago, and "non-Neanderthal" to be all humans alive in Africa 100,000 years ago, we can unambiguously identify the set of people who are 100% non-Neanderthal and the set of people who are 100% Neanderthal. Similarly, if we are talking about recent admixture in African Americans, we might define as "European" everyone who lived physically in Europe 1,000 years ago and "African" everyone who lived physically in Africa 1,000 years ago. The critical issue here is that the definition of a population is operational and has a temporal aspect that depends on the respective context. By considering the definitions within a particular context, the assumptions and limitations of these definitions can be made clear.[19]

Sociologists Aaron Panofsky and Catherine Bliss have shown empirically that genomics researchers are ambiguous when it comes to defining populations, and they have argued that this ambiguity may actually be

[18] M'charek, A. (2005). *The Human Genome Diversity Project: An ethnography of scientific practice.* Cambridge: Cambridge University Press, pp. 33–42.

[19] I am indebted to Iain Mathieson for this example and for pointing out this legitimate use of the concept of admixture in population genetics.

154 ANCESTRY REIMAGINED

intentional. Panofsky and Bliss performed a content analysis of the articles presenting original research in three yearly issues of the journal *Nature Genetics* (1993, 2001, and 2009), with the aim to document the population labels used in the articles. They thus identified eight systems for classifying populations: race (e.g., White or Caucasian); continent (e.g., African); continental region (e.g., Northern European); country (e.g., Japanese); country region (e.g., Sicilian); ethnicity (e.g., Han); language (e.g., Bantu speakers); and other (usually religion, e.g., Jewish). Panofsky and Bliss recognized that these eight schemes correspond to three kinds of identity and belonging: racial (the first two systems), ethnic (the last three systems), and geographical (the remaining three systems). The content analysis showed that whereas in 1993 half of the articles did not specifically mention human populations, in 2009 most articles did. A second finding was that racial classification decreased across time, with the most significant increase in the use of continental labels. This resulted in ambiguity, as Panofsky and Bliss observed, because some continental labels such as "African" also have racial connotations. However, country labels were the ones most frequently used, in contrast to ethnic, linguistic, and religious labels that were less frequently used. Perhaps the most important finding was that geneticists mixed these classification systems: the same label, for instance "East Asian," might be used in the same article as referring to race in one place and as referring to geography in another place. Therefore, whereas in 1993 and 2001 about 60% of the articles used a single classification system and were thus consistent about what a population was considered to be (a racial group, a geographic group, or an ethnic group), in 2009 this was the case for 40% of articles (and some of these also used ambiguous classifications such as "East Asian"). As Panofsky and Bliss concluded, "Clarification, purification, and standardization of classification practices are not the trend. Rather, combination, hybridization, and ambiguity in human population classification—classificatory polyvocality— have increased over time."[20]

Even the way that reference is made in different studies to what is usually considered to be the same population can be confusing. For instance, people of European ancestry living in the United States have been designated in a variety of ways in scientific articles, such as:

[20] Panofsky, A., and Bliss, C. (2017). Ambiguity and scientific authority: Population classification in genomic science. *American Sociological Review, 82*(1), 59–87.

SOCIAL CONSTRUCTS VERSUS "NATURAL ORDER"   155

- *European Americans*: This term is used to distinguish this group from Native, African, and Japanese Americans, among others. However, this group can include people from several different parts of Europe such as Russia and Ireland, or Finland and Italy. Therefore, their description as European Americans masks this diversity.
- *White people or Whites*: This term is supposed to describe the skin color of people who come from Europe. However, skin color is not always indicative of a person's DNA and ancestry, as several people with white skin color in the United States have been found to have African ancestry.
- *Caucasoids or Caucasians*: This is a term that as we saw was used in the attempt to distinguish between what, during the 19th and the 20th century, was supposed to be distinct human races (Chapter 3). The term was supposed to indicate origin from Europe, and it persisted because the skull that best represented their characteristics had been found in the Caucasus Mountains in Eastern Europe.[21]

According to public health geneticist Alice Popejoy, seven out of ten clinical genetics laboratories in the United States that share the most data with researchers use the term "Caucasian" as a choice to designate the patients' racial or ethnic identity; in addition, almost 5,000 scientific papers in biomedicine since 2010 have used this term to describe European populations. As she pointed out, "This suggests that too many scientists apply the term, either unbothered by or unaware of its roots in racist taxonomies used to justify slavery—or worse, adding to pseudoscientific claims of white biological superiority."[22] In other words, the terms we use have a history, and it is important to keep that in mind.

Recently, there have been several calls from geneticists to reconsider the language used in population genetics studies. The main point has been that they should be more concerned about the terms used, and about how those might be interpreted. Therefore, a careful choice of terms is necessary in order to avoid misleading interpretations of the findings. For instance, in a paper titled "The Language of Race, Ethnicity, and Ancestry in Human Genetic Research," geneticists Ewan Birney, Michael Inouye, Jennifer Raff,

---

[21] Jobling, M., Hollox, E., Hurles, M., Kivisild, T., and Tyler-Smith, C. (2013). *Human evolutionary genetics* (2nd ed.). New York: Garland Science, p. 5.
[22] Popejoy, A. B. (2021). Too many scientists still say Caucasian. *Nature, 596*(7873), 463.

Adam Rutherford, and Aylwyn Scally suggested (and I quote because I could not agree more):

> We argue that a more critical understanding by geneticists of the societal impact of their work and how it is communicated is necessary in order to minimise or prevent misunderstanding. A practical implication of this is the need for new concepts, terminology and language to enable scientists to communicate accurately and appropriately both within their field and to other scientific and lay audiences. Humans are social creatures, and it is no surprise that social interactions have profound effects on many human phenotypes, sometimes outweighing the effect of an individual's own genetic make-up. Although often the goal in studying genomes is to find genetic causes of phenotypic variation, in practice this tends to minimize other sources of variation outside of genetics, including social variation. In particular, we call for an approach to describing methods and results which recognizes the large effects of non-genetic factors—particularly social, cultural, and economic—on human traits and uses language that reflects the entangled nature of culture and genetics in human life. There are other fields—anthropology, psychology and sociology—with extensive histories of describing the impacts of interpersonal behaviours in family, educational and cultural settings, and we advocate for a deeper engagement with researchers from these fields—ideally through sustained research collaboration—in helping shape the terminology we use and even the design of our studies.[23]

Defining and individuating human populations is abstract, because human DNA variation is continuous and there is thus not sufficient DNA diversity for differentiation. Let us see why.

Clines, Not Clusters

I have already mentioned that human DNA variation is clinal, due to particular historical processes. One is called isolation by distance. Within each population more individuals move to nearby places and fewer travel to places

---

[23] Birney, E., Inouye, M., Raff, J., Rutherford, A., and Scally, A. (2021). The language of race, ethnicity, and ancestry in human genetic research. arXiv preprint: 2106.10041.

far away. Therefore, most individuals in a population tend to mate with other individuals from the same or neighboring populations. As a result, their descendants can inherit the DNA variants that exist in their own or in their neighboring populations, and not those DNA variants that may exist in more distant ones. In this way, the DNA diversity in population A will gradually decline the further away one moves from the area in which population A is located, because fewer and fewer of the DNA variants in population A will exist in more distant populations (because the people therein and those in population A are less likely to mate). The model of isolation by distance entails that the closer two populations are geographically, the more similar at the level of DNA they will be. This model explains much of the general continuous pattern of human DNA variation that is characterized by clines, that is, gradients of DNA variant frequencies, as in the Novembre and colleagues' study of human DNA variation in Europe discussed in the previous chapter.

Clinal genetic variation can also be the outcome of migratory processes of humans and the related founder effects: when populations migrate, some individuals tend to settle in a region and only a subset of them is expected to migrate farther away. As a result, only a subset of the genetic variation of the initial population is represented in the group of individuals who migrated. For instance, imagine that in an initial population B, there exist various alleles, say 30, in various frequencies. When a small group of individuals of this populations migrates, it will consist of a random sample of individuals with respect to these alleles. Therefore, it is likely that the first group of migrants might have 20 of the 30 alleles, due to them being a random sample of the initial population. If these individuals establish a new population B* of similar size with the original one, under certain conditions it will have fewer alleles, and therefore less DNA diversity, than the initial population B. A new migratory event starting from population B* could end up producing a new population B**, which might have an even lower DNA diversity than both B and B*, because it would be less likely for individuals migrating to carry all 20 alleles of B*. This could eventually produce a cline of genetic variation in which DNA diversity would decrease with geographic distance from the location of population B toward the location of population B**.

Finally, in some cases, clinal genetic variation reflects natural selection and adaptation because of the clinal variation in environmental factors due to which populations adapt. Perhaps the most characteristic example in humans is skin color, already discussed in Chapter 3 (see Boxes 3.1 and 3.2). This is actually the product of two environmental clines. The first cline relates

158 ANCESTRY REIMAGINED

to protection from high levels of ultraviolet radiation (UV), which is higher the closer an area is to the equator and lower the further away from it. The second cline relates to the production of vitamin D in low-UV environments far from the equator. As human populations dispersed out of Africa, they experienced different intensities of UV radiation. Because there are high levels of UV radiation near the equator, natural selection favored dark pigmentation in those places. Away from the equator, and toward the two poles, UV levels are generally low, and so natural selection favored the evolution of depigmented skin (that is, having lighter skin color), which allows UV photons to sustain synthesis of vitamin D.[24]

It therefore seems that there are several reasons for human genetic variation to be clinal. If this is the case, then how it is possible to observe clusters where clines exist? This can happen when the samples compared are not representative of the extant variation, a point already raised by Pääbo and Serre in the previous chapter. Imagine an area divided in five zones of genetic variation (1–5). Imagine also that there is clinal variation, that is, a gradient in variation of particular DNA sites. For convenience, let's think of this genetic variation as being represented by the variation in color (and hairstyle; see Figure 8.1a). In this particular case, variation is clinal: the frequencies decrease or increase (depending on how you look at it) across geographic locations. Depending on which zones researchers will manage to acquire samples from, the conclusions about the genetic differences of the compared populations might be significantly different. If researchers acquired samples only from zones 1 and 5, their conclusion would be that two distinct populations exist that differ in the variant(s) of interest (8.1b.i). However, if they managed to acquire samples of zones 1, 3, and 5, the conclusion would be different as they might be able to realize that there are no distinct populations but rather a continuum of variation (8.1b.ii). This would become even clearer if they managed to sample all the available variation from zones 1–5 (8.1b.iii), something that is usually very difficult, if not impossible in practice. What is important to note is that the differences among the various groups in zones 1–5 are differences in degree, not kind. These groups are parts of a cline, a continuum of variation, where it is not possible to find absolute and strict boundaries. Clusters can thus be found due to sampling bias (8.1b.i), because human DNA variation is clinal.

---

[24] Jablonski, N. G., and Chaplin, G. (2010). Human skin pigmentation as an adaptation to UV radiation. *Proceedings of the National Academy of Sciences, 107*(Supplement 2), 8962–8968.

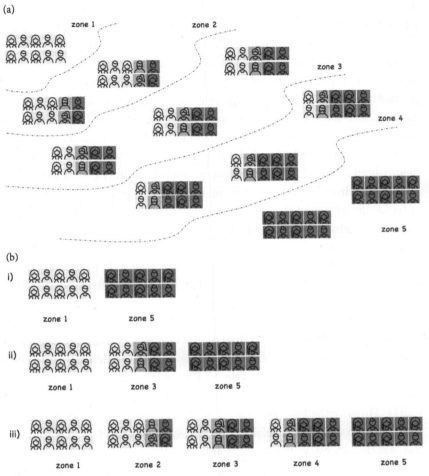

Figure 8.1 How nonrepresentative sampling can give the impression of genetically clustered groups where clines of genetic variation actually exist. Whereas the actual clinal genetic variation in an area is the one shown in (a), in (b) it is shown that nonrepresentative sampling, that is, sampling from some but not all of the areas can give the impression of clusters (b.i), whereas clines actually exist (b.iii). (Free Icons from the Streamline Icons Pack)

How is this established? A study aimed at assessing human DNA variation with a more representative sampling procedure. Researchers thus combined data from already available samples from previous studies, such as the HapMap Project, with data from 296 individuals from 13 worldwide populations, including populations from West Africa, Central Europe, West

160   ANCESTRY REIMAGINED

Asia, Central Asia, South Asia, Southeast Asia, Polynesia, and America, which had not been covered in previous studies. This resulted in a dataset of 850 individuals from 40 populations. The researchers compared the population differentiation among Africa, Asia, and Europe in the following datasets: (1) the HapMap data alone (210 individuals from four HapMap populations YRI, CEU, CHB, and JPT—see Chapter 6); (2) the HapMap data and a dataset of 344 individuals from 23 populations from a previous study of theirs,[25] and (3) both datasets in (2) and the new dataset from the not previously covered 13 populations. The results of their analysis showed that "differentiation among human populations decreases substantially and genetic diversity is distributed in a more clinal pattern when more geographically intermediate populations are sampled." This is exactly what is shown in Figure 8.1. The researchers pointed out that human DNA variation can be influenced by geographic factors, such as mountains, and cultural factors, such as language and religious practices, and so in particular cases it is not perfectly clinal. But they also noted: "However, between-population differences can be seriously exaggerated if human populations are sparsely sampled."[26]

Additional evidence for the clinal nature of human variation comes from a recent study of BioMe, a large and highly diverse multiethnic biobank in New York City. Among these people, 30,376 reported only one race/ethnicity category as follows: 9,830 European American, 7,976 African American, 11,544 Hispanic/Latino, 965 East and Southeast Asian, and 61 Native American. To explore the relationship between self-reported race/ethnicity and genetic ancestry, the researchers estimated global genetic ancestry proportions for 31,705 BioMe participants. Using principal components analysis (PCA), the researchers showed that BioMe participants represent a continuum of genetic diversity (Figure 8.2). The details in this figure are not necessary to understand. What matters is that the various participants who are colored based on their self-reported ancestry did not form genetically distinct clusters but fell along a continuum of genetic variation.[27] This has

[25] Xing, J., Watkins, W. S., Witherspoon, D. J., Zhang, Y., Guthery, S. L., Thara, R., Mowry, B. J., Bulayeva, K., Weiss, R. B., and Jorde, L. B. (2009). Fine-scaled human genetic structure revealed by SNP microarrays *Genome Research, 19,* 815–825.

[26] Xing, J., Watkins, W. S., Shlien, A., Walker, E., Huff, C. D., Witherspoon, D. J., et al. (2010). Toward a more uniform sampling of human genetic diversity: A survey of worldwide populations by high-density genotyping. *Genomics, 96*(4), 199–210.

[27] Belbin, G. M., Cullina, S., Wenric, S., Soper, E. R., Glicksberg, B. S., Torre, D., et al. (2021). Toward a fine-scale population health monitoring system. *Cell, 184*(8), 2068–2083.

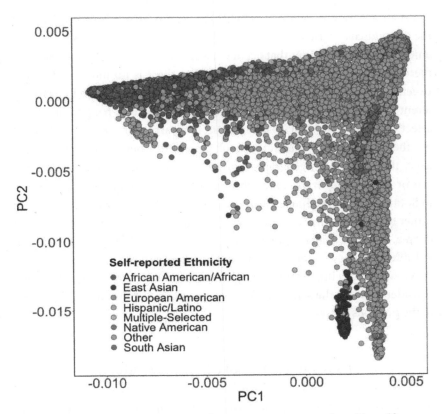

Figure 8.2 PCA for 31,705 genotyped BioMe participants colored by self-reported race/ethnicity. Details notwithstanding, it is evident from this database that there are no distinct groups; rather human DNA variation is continuous. (Reprinted with permission from Belbin, G. M., Cullina, S., Wenric, S., Soper, E. R., Glicksberg, B. S., Torre, D., et al. (2021). Toward a fine-scale population health monitoring system. *Cell*, 184(8), 2068–2083.)

prompted scholars to argue that researchers should refrain from using continental ancestry categories, which can be easily confused with racial groups, and embrace a multidimensional, continuous view of ancestry.[28] As shown in Figure 8.2, human DNA variation is continuous—this continuity is all you need to grasp from this figure.

---

[28] Lewis, A. C., Molina, S. J., Appelbaum, P. S., Dauda, B., Di Rienzo, A., Fuentes, A., et al. (2022). Getting genetic ancestry right for science and society. *Science*, 376(6590), 250–252.

162 ANCESTRY REIMAGINED

When geographic and cultural barriers limit or preclude mating among the individuals of two neighboring populations, any new DNA variants emerging from mutations that occur in each one of them remained restricted there and are not transferred to the other as there is no interbreeding between their members. As a result, these populations can evolve to have different frequencies for particular genetic variants. This also happens if natural selection favors a particular DNA variant in a particular population, but not in others where no such selection took place. But wherever genetic clusters exist, they depend on geography and do not necessarily correlate with racial or ethnic categories, although in some cases they might; for example, in South Asia, some genetic clusters correspond to caste or religious groups rather than geography.[29] Furthermore, the difference between them is one of degree, not kind: as already mentioned, there exist no SNPs that can be found at 100% frequency in one area only, and nowhere else.[30]

As sampling is important for studying human DNA variation, let us then consider in detail the sampling procedures of some of the studies discussed in the previous chapter.

How Sampling Can Affect the Findings

As already mentioned, the most usual way to designate a population in population genetics studies is to let participants identify themselves with a population. For instance, in the study of the genetic diversity in Europe by Novembre and colleagues discussed in the previous chapter, the inclusion or exclusion of participants was based on their presumed geographic ancestry. Whenever this was not possible to do, the researchers assigned a geographic origin of participants based on the origin of their grandparents: if all of them had originated from a single country, that was considered as their country of origin. In contrast, those individuals whose grandparents came from different countries were excluded. In cases for which information about the country of origin of the grandparents was not available, they used the

[29] Reich, D., Thangaraj, K., Patterson, N., Price, A. L., and Singh, L. (2009). Reconstructing Indian population history. *Nature, 461*(7263), 489–494.

[30] Fujimura, J. H., Bolnick, D. A., Rajagopalan, R., Kaufman, J. S., Lewontin, R. C., Duster, T. et al. (2014). Clines without classes: How to make sense of human variation. *Sociological Theory, 32*(3), 208–227; Feldman, M., and Lewontin, R. C. (2008). Race, ancestry, and medicine. In B. Koenig, S. Lee, and S. Richardson (Eds.), *Revisiting race in a genomic age.* New Brunswick, NJ: Rutgers University Press, pp. 89–101.

SOCIAL CONSTRUCTS VERSUS "NATURAL ORDER"   163

individual's country of birth as the country of origin. The researchers also excluded individuals whose geographic origin was outside of Europe, those who might be related to one another, and those who were found to be distinct from the others (described as outliers) in a preliminary PCA. Remember also that samples were collected in two places only, Lausanne and London.[31] There are at least two kinds of issues with this sampling method.

The first issue concerns the people who were included in the study. Considering that an individual has an origin in a particular country insofar as their grandparents all came from that country may sound reasonable, but it can actually be misleading. Take my case as an example. I was born and grew up in Greece, so I am considered and consider myself to be Greek—no surprise here. All my grandparents were born in Greece in the early 20th century, and they all lived their whole lives there. This was also the case for my great-grandparents (I came to know three of them, and I know about the others from my parents). Therefore, I have a good, personal sense of my family tree up to three generations back—about people born in the late 19th or early 20th century. However, I personally have no idea whether the ancestors of my great-grandparents were also born in Greece. I know for a fact that none of my grandparents had any connection to the more than 1 million people who migrated to Greece from Asia Minor after the catastrophic events of 1922 and the population exchange of 1923.[32] I know for a fact that none of my grandparents or great-grandparents ever lived in Asia Minor or spoke Turkish. However, I do not know if any ancestor of theirs originated in that area, or elsewhere. Many Greeks, or Greek-speaking people, lived in Asia Minor for hundreds of years before the Greek revolution of 1821, and I do not know if any of my ancestors had an origin from the area that today is Turkey. Therefore, having all of a person's grandparents born in the same country may be more than sufficient for that person to identify with that particular ethnic identity, but it is a bit short-sighted when it comes to genetic ancestry.

---

[31] Novembre, J., Johnson, T., Bryc, K., Kutalik, Z., Boyko, A. R., Auton, A., et al. (2008). Genes mirror geography within Europe. *Nature, 456*(7218), 98–101.

[32] About 1,100,000 people moved to Greece from Asia Minor, and another 100,000 from Russia and Bulgaria, whereas about 380,000 people returned to Turkey—the criterion for the exchange was religion, rather than language or "national consciousness"; as a result, many Christian Orthodox people who were sent to Greece spoke Turkish, and many Muslims who returned to Turkey, especially those from Crete, spoke Greek. See Clogg, R. (2021). *A concise history of Greece* (4th ed.). Cambridge: Cambridge University Press, pp. 98–101.

164   ANCESTRY REIMAGINED

The second issue concerns the people who were excluded from the study. In their supplementary materials, Novembre and colleagues acknowledged that "the rule of using reported grandparental origins as a proxy for genetic ancestry can in rare cases be misleading," mentioning the following examples: an individual with four Italian grandparents, who was born in France, spoke French, and clustered with French individuals; an individual with four Russian grandparents, who was born in Romania, spoke Romanian, and was placed between Switzerland and Romania; an individual with four Swiss grandparents, who was born in Italy, spoke Italian, and clustered with Italian individuals; an individual with four Swiss grandparents, who was born in Israel, had parents who are Italian, spoke Italian, and clustered with Italians; and an individual who had four German grandparents and who was born in Hungary, spoke Hungarian, and clustered with Italian individuals. These were considered as outliers and were not included in the analysis, along with those people who had grandparents born in different countries, because they could not be assigned to a single country.

Table 8.1 presents the sampling procedure that Novembre and colleagues followed (included in the supplementary materials of their article). From the initial sample of 3,192 people, 1,805 were excluded for various reasons. This means that more than half of the initial sample (56%) was excluded from the analysis! This has been criticized by molecular evolutionist Robert DeSalle and paleoanthropologist Ian Tattersall as cherry-piking of data: "when you cherry-pick a data set you will obtain a neat answer that agrees well with

Table 8.1 Exclusion Criteria in the Novembre et al. 2008 Study

Exclusions	Stage of Analysis	Sample Size
	Total individuals of European descent	3,192
259	After exclusion of individuals with origins outside of Europe	2,933
524	After exclusion of individuals with mixed grandparental ancestry	2,409
24	After exclusion of putative related individuals	2,385
34	After exclusion based on preliminary PCA run	2,351
964	After thinning Swiss-French and UK individuals	1,387
1,805	Total excluded	

SOCIAL CONSTRUCTS VERSUS "NATURAL ORDER"   165

the rules under which you did your cherry-picking."[33] I must also note that among the 1,387 individuals included in the analysis, it was possible to confirm the country of origin of all of each individual's grandparents for 771 individuals. For the remaining individuals this information was not available, and so the self-reported country of birth was used for this purpose. This means that for more than half of the sample we do not know if some individuals should have been excluded according to the criteria in Table 8.1, had the information about the origin of all four grandparents been available.

Taking these exclusion criteria into account is important, because they affect the sizes and the representativeness of the samples used in analysis. In his landmark 1972 study of human genetic diversity,[34] evolutionary geneticist Richard Lewontin asked two critical questions about the sampling process:

- Representativeness: How many nationalities should one include from the same continent?
- Weighting: Should subsamples be weighted; that is, should one consider the proportion of the total population that they represent?

With respect to the study by Novembre and colleagues that we are considering here, one can thus ask accordingly how many subpopulations from the same country should have been included and whether the samples should have been weighted in order to consider the proportion of the total population they represent. None of this was done, even though, as shown in Table 8.1, the researchers removed 964 individuals from the United Kingdom and the Swiss-French group to achieve a more balanced representation, as they likely had many people from these groups due to the locations of sample collection (London and Lausanne, respectively). Table 8.2 presents the sample sizes for each country used in this study (included in the supplementary materials of their article).

Let us consider representativeness first. The participants in the study were identified only on the basis of their country of grandparental origin, without knowing from which specific part of that country they may have originated. This is a lack of resolution regarding geography. As a result, samples may not be representative of the respective populations they are supposed to

[33] DeSalle, R., and Tattersall, I. (2018). *Troublesome science: The misuse of genetics and genomics in understanding race*. New York: Columbia University Press, p. 93.

[34] Lewontin, R. C. (1972). The apportionment of human diversity. *Evolutionary Biology*, 6, 381–398.

166 ANCESTRY REIMAGINED

Table 8.2 Summary of Sample Sizes Used for Each Country in the Novembre et al. 2008 Study

Geographic Origin	No. of Individuals	Geographic Origin	No. of Individuals
Italy	219	Bosnia and Herzegovina	9
United Kingdom	200	Croatia	8
Spain	136	Greece	8
Portugal	128	Russian Federation	6
Swiss-French	125	Scotland	5
France	91	Cyprus	4
Swiss-German	84	Macedonia	4
Germany	71	Turkey	4
Ireland	61	Albania	3
Serbia and Montenegro	44	Norway	3
Belgium	43	Bulgaria	2
Poland	22	Kosovo	2
Hungary	19	Slovenia	2
Netherlands	17	Denmark	1
Austria	14	Finland	1
Romania	14	Latvia	1
Swiss-Italian	13	Slovakia	1
Czech Republic	11	Ukraine	1
Sweden	10		

represent because of sampling bias: some people from specific regions of a country happened to participate, whereas others did not.[35] The only country for which this was done was Switzerland, probably because one of the two collection sites was there (Lausanne). The analysis involved 125 individuals from the French-speaking part, 84 individuals from the German-speaking part, and 13 individuals from the Italian-speaking part, and the researchers claimed that they were able to clearly distinguish among these three subpopulations. What would the outcome of the analysis have been had the researchers done the same for all countries? If they were able to differentiate

[35] Geary, P. J., and Veeramah, K. (2016). Mapping European population movement through genomic research. *Medieval Worlds*, 4, 65–78.

among the inhabitants of a small country such as Switzerland, wouldn't they also be able to do the same had they considered subpopulations from larger countries such as France, Germany, and Italy?

This brings us to the issue of weighting. Does it make sense to have data from 222 individuals for Switzerland, which was a country of 7.702 million residents in 2008,[36] and 91 individuals from France, which is a country of 63.961 million residents in 2008 (this was when the Novembre and colleagues' study under discussion was published)?[37] The fact that the researchers removed 964 individuals from the United Kingdom and the French-speaking part of Switzerland does not resolve this issue. If the analysis were able to distinguish among people coming, say, from Geneva (Swiss-French part), Zurich (Swiss-German part), and Lugano (Swiss-Italian part), wouldn't it also be able to differentiate among people coming, say, from Strasbourg, Marseilles, and Bordeaux, the geographic distance among which is a lot larger than among the aforementioned Swiss cities? This would obviously be the case for any other country. Such a sampling bias could be addressed by subsampling or resampling the dataset in order to "examine the robustness of ancestry inference results to variation in sampling."[38]

In their article, Novembre and colleagues noted that their analysis did not yield any separate clusters within Europe as it would have been expected for discrete and well-differentiated populations, whereas in the studies by Rosenberg and colleagues discussed in the previous chapter such clusters emerged for continental regions. One might conclude that Europeans are very similar genetically to one another so that distinct clusters cannot be observed, whereas this is not the case when we compare Europeans to other continental groups. This sounds reasonable, until one considers how the sampling procedure may affect the outcome of the analysis. It is only due to sampling bias that it is possible to distinguish between neighboring populations if human genetic variation is clinal. If you are wondering how sampling was done in the various studies discussed in the previous chapters, take a look at Figure 8.3, which shows the geographical distribution of the 52 populations included in the HGDP and which were used in the analyses presented in Chapter 7 that yielded clusters that corresponded to continental groups (by Rosenberg and colleagues, and Li and colleagues). As is obvious

---

[36] https://www.bfs.admin.ch/bfs/en/home/statistics/population.html (accessed January 29, 2022).
[37] https://www.insee.fr/en/statistiques/2382601?sommaire=2382613 (accessed January 29, 2022).
[38] Shringarpure, S., and Xing, E. P. (2014). Effects of sample selection bias on the accuracy of population structure and ancestry inference. *G3: Genes, Genomes, Genetics*, 4(5), 901–911.

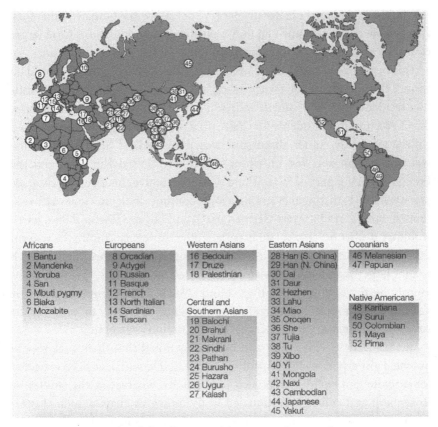

Figure 8.3 Geographical distribution of the 52 populations that were represented in the samples of the Human Genome Diversity Project. (Reprinted with permission from Cavalli-Sforza, L. (2005). The Human Genome Diversity Project: past, present and future. *Nature Reviews Genetics*, 6, 333–340.)

in Figure 8.3, in many cases the actual sampling seems to be closer to the model presented in Figures 8.1.b.i and 8.1.b.ii, rather than the one in Figure 8.1.b.iii. This means that there were whole areas that are not represented in the sample.

This is a problem that Cavalli-Sforza himself acknowledged in the article from which the map in Figure 8.3 comes. He wrote that the main limitation of the HGDP collection available at the time was that both the number of populations and the number of individual samples were small. As the populations did not "evenly cover the inhabited regions of the world," he suggested that the main future requirement would be to increase the number

SOCIAL CONSTRUCTS VERSUS "NATURAL ORDER"   169

of samples, particularly from insufficiently represented areas. He acknowledged that areas such as India and Polynesia were not represented at all, whereas Europe, northern Asia, the Americas, and Oceania had limited representation. He also acknowledged that whereas the average population sample was 50 individuals, the sample size of each population varied between 1 and 50 individuals, while large populations such those from China and Pakistan were represented by a sample of 10 individuals only.[39] Once again, here is a case where scientists themselves are aware of the limitations and are explicit about those. I should note though that Cavalli-Sforza did not express any concern about the findings of the studies that relied on the HGDP data. However, this is not the only problem, and perhaps not even the most important one, as additional sampling might resolve any issues.

There is another problem, which in my view is a lot more important and that has to do with the choice of populations. As Cavalli-Sforza explained for the HGDP, "all samples are from populations of anthropological interest— that is, those that were in place before the great diasporas started in the fifteenth and sixteenth centuries, when navigation of the oceans became possible. This choice was important, because these diasporas caused significant population admixtures, especially in the Americas but also in other continents," defining *admixture* in the same article as "the mixture of two or more genetically distinct populations." Cavalli-Sforza concluded that "Only genetic knowledge of the original populations that contributed to these admixtures can disentangle the various genetic complexities that resulted, and the HGDP fulfils these criteria."[40] This view thus assumes that before the 15th century there were no population movements around the world, and as a result, any populations living far from each other were genetically distinct. When European colonization, and the subsequent slave trade, began, many populations became admixed because of large population movements. Therefore, Cavalli-Sforza argued that it was necessary to include in the project "well-chosen populations," for which researchers would be certain that they were unadmixed and thus genetically distinct from one another. The comparison, therefore, of these unadmixed ("pure") populations was expected by Cavalli-Sforza to yield important information about population movements. The assumption was that the chosen populations were

[39] Cavalli-Sforza, L. (2005). The Human Genome Diversity Project: Past, present and future. *Nature Reviews Genetics*, 6, 333–340.
[40] Cavalli-Sforza, L. (2005). The Human Genome Diversity Project: Past, present and future. *Nature Reviews Genetics*, 6, 333–340.

170 ANCESTRY REIMAGINED

living representatives of the original population before the colonialism era. Historian Marianne Sommer aptly described this view as "the conceptualization of indigenous peoples as archives of historical documents that is associated with a certain nostalgia of purity, the imagining of a world lost in time when groups kept to themselves and carried clean signs of unequivocal kinship in their genomes."[41]

However, this assumption is wrong. There is ample archaeological evidence that humans were migrating around the globe before the 15th century (they were not circumnavigating, of course, but neither were they entirely static). Indeed, we might argue that globalization, defined as a complex connectivity due to a dense network of intense interactions and interdependencies between disparate people brought together by the long-distance flow of goods, ideas, and people, is not a recent phenomenon. Of course, the pace and the extent to which this network has developed during the last 500 years or so are unparalleled. Yet this does not mean that globalization did not occur in the past. In fact, some scholars have argued that throughout human history there has been a single trend toward increasing globalization. Others have argued that globalization has occurred repeatedly in the past for certain periods of time. Whatever the case, the important point is that trends associated with globalization, such as those we observe today, can be found in earlier eras as well.[42] What seems to be more important though, is not the idea of "global" but the idea of "complex connectivity," as was the case for ancient Rome, for instance. However, one might wonder, what is the point of talking about globalization if the "global" aspect is not to be taken literally? The answer is that globalization refers not to the idea that everything has to be connected, but to the idea that everything could be connected. So what matters for our discussion here is that even if interactions were not literally as global as they are today, still complex connectivity and intense interactions and interdependencies could be observed between disparate groups of people.[43] In such a case, it makes no sense to talk about unadmixed populations of the past.

---

[41] Sommer, M. (2016). *History within: The science, culture, and politics of bones, organisms, and molecules.* Chicago: University of Chicago Press, p. 326.

[42] Jennings, J. (2017). Distinguishing past globalizations In T. Hodos (Ed.), *The Routledge handbook of archaeology and globalization* (pp. 12–28). New York: Routledge.

[43] Knappett, C. (2017). Globalization, connectivities and networks: An archaeological perspective. In T. Hodos (Ed.), *The Routledge handbook of archaeology and globalization* (pp. 29–41). New York: Routledge.

SOCIAL CONSTRUCTS VERSUS "NATURAL ORDER"   171

A last limitation to consider, already discussed, is that the analysis of DNA data from people who live today cannot provide any information about what happened in the past beyond the most recent events, or about where ancestral populations have lived because human populations have not been static. The fact that particular people live today in an area does not entail that they are direct descendants of ancestral populations who lived there in the past. This is why terms such as "Indigenous" or "native" are relative and should be used with caution (see Chapter 2). But even if modern populations are representative of past ones, current genetic variation does not necessarily represent the variation of past populations. This happens not only because some individuals but not others in each generation produced offspring, but also because DNA inheritance is diluted across generations (see Chapter 4) and so only some ancestral DNA segments will be found in people living today. To this we should add the issue with the dating of past events based on DNA. Because rates of mutation across the human genome and between individuals may vary, any estimate made will depend on the method used. If the currently accepted mutation rate is revised in the future, any dating of past events in already published studies will also be affected.[44]

Given all these limitations, one should be careful about what inferences can be made about the past from what we presently know. The analyses of DNA can be reliable and valid, but their interpretation in terms of history requires extreme caution. What has to be avoided are also superficial assumptions about how the peoples of Europe, and those of other regions, came to exist. The limitations we considered here should be kept in mind because studies of human DNA can be misinterpreted as providing the grounds for divisive and discriminatory attitudes (and sometimes policies) of racial, national, or ethnic belonging or for the perceiving races, nationalities, or ethnicities as having a genetic basis. None of this is the case, and so any distinctions between "us" and "them" groups has no natural basis on DNA. As medieval historian Patrick Geary nicely put it, we should be concerned by the assumption:

that the peoples of Europe are distinct, stable and objectively identifiable social and cultural units, and that they are distinguished by language,

---

[44] Geary, P. J., and K. Veeramah (2016). Mapping European population movement through genomic research. *Medieval Worlds*, 4, 65–78; Nielsen, R., Akey, J. M., Jakobsson, M., Pritchard, J. K., Tishkoff, S., and Willerslev, E. (2017). Tracing the peopling of the world through genomics. *Nature*, 541(7637), 302–310.

172 ANCESTRY REIMAGINED

religion, custom, and national character, which are unambiguous and immutable. These peoples were supposedly formed either in some impossibly remote moment of prehistory, or else the process of ethnogenesis took place at some moment during the Middle Ages, but then ended for all time. . . . Europe's peoples have always been far more fluid, complex, and dynamic than the imaginings of modern nationalists. Names of peoples may seem familiar after a thousand years, but the social, cultural, and political realities covered by these names were radically different from what they are today.[45]

The groupings used in the population genetics studies are human inventions, and this is something that should always be kept in mind. The reason for this is not only that the criteria for inclusion and exclusion are subjective, and sometimes biased; it is also that it is practically difficult to achieve a sampling that accurately represents the extant human DNA variation. Therefore, any patterns we might find always depend on the samples, as well as the methods used for the analyses. These methods are robust and reliable, but what they provide is a pattern that we should explore further, not any definitive answers for the relatedness and the genetic history of these populations.

But if modern DNA, that is, DNA from people living today, cannot provide information about past populations, perhaps ancient DNA can? It is to this topic to which we now turn.

---

[45] Geary, P. J. (2002). *The myth of nations: The medieval origins of Europe.* Princeton, NJ: Princeton University Press, pp. 11, 13.

9

Separating DNA from Culture

Ancient DNA

Ancient DNA (aDNA) is DNA that has been found in ancient specimens, in contrast to modern DNA that is found in people living today. But how old is "ancient"? The answer is that it depends, as the term is used for DNA found in remains of a few hundred years ago, as well as for those of several thousand years ago as in the case of Neanderthals and the Denisovans. A pioneering figure in this area of research is geneticist Svante Pääbo, who we met already in Chapter 7 and who in the 1980s was among the first to show that DNA can be retrieved from the remains of dead individuals. Before this, geneticists were obliged to work with modern DNA only. However, once the study of aDNA became possible, this "time trap" was overcome, and it made feasible what Pääbo himself and his colleague Johannes Krause described as "genetic time travel." This was not achieved at once, though, but only when next generation sequencing (NGS) became available (see Box 1.2).[1] But the study of aDNA is not without challenges! In fact, it is a lot more difficult to study aDNA than modern DNA. There are at least three reasons for this.

The first one is that aDNA is not easy to find, as it does not exist in fossils. Fossils are the remains of the hard parts of organisms (usually the skeletons) that have been preserved, maintaining the organisms' original shapes. What has happened in this case is that the organic material in the tissues has been replaced by minerals. Researchers need to find preserved tissues, usually teeth or bones, that have not been completely mineralized, so that there is a chance that DNA could be extracted from those. If one finds fossils, there will be no DNA there.[2]

---

[1] Krause, J., and Pääbo, S. (2016). Genetic time travel. *Genetics, 203*(1), 9–12. aDNA has specific features: (a) most aDNA molecules, especially those from the chromosomes, are around 50 bp long, and thus cannot be multiplied using PCR but can be studied with NGS; (b) aDNA molecules can be detected with NGS (after death, DNA molecules undergo damage and scientists can characterize the damage patterns), and thus help distinguish them from modern DNA molecules.

[2] Reich, D. (2018). *Who we are and how we got here: Ancient DNA and the new science of the human past.* Oxford: Oxford University Press, p. 31.

Ancestry Reimagined. Kostas Kampourakis, Oxford University Press. © Oxford University Press 2023.
DOI: 10.1093/oso/9780197656341.003.0009

174 ANCESTRY REIMAGINED

The second reason is that aDNA can be damaged across time, so that even if we find some it will not be intact. While organisms are alive, chemical modifications in the structure of DNA can be "repaired" by enzymes. As a result, the integrity of DNA molecules can be maintained. But after an organism dies, the repairing mechanisms of cells stop functioning. This can result in the destruction of DNA due to cellular enzymes that are no longer restrained and can access it, due to microorganisms who in the meantime propagate in the dead tissues, or due to the damaging effects of water and oxygen that chemically react with it. However, if tissues become frozen or become desiccated immediately after death, these processes that destroy DNA are inhibited. In such cases, it is possible to isolate and analyze DNA that has been preserved in the tissue. One major problem with aDNA is its degradation (or fragmentation), which is actually the reason why the molecules of aDNA usually found are very short (around 50 bp). Because of this degradation, it is possible that the DNA segments found in ancient bones and analyzed are incomplete, lacking parts that would be necessary for comparisons and accurate conclusions. There are also chemical modifications that DNA can undergo, which prevent their amplification and sequencing, or lead to their "misreading."[3]

Another obstacle in the study of aDNA is its contamination with modern DNA, coming from bacteria and fungi from the immediate environment that have decomposed the tissues of dead people, or from the humans who handled the specimens from which aDNA was derived. Especially in the latter case, DNA analysis can be very misleading. A typical Neanderthal DNA fragment from a well-preserved sample is only 40 base-pairs long, whereas the average difference between modern humans and Neanderthals is about 1/600. Therefore, in some cases there will be no differences in the DNA sequences of humans and Neanderthals, and it will thus be impossible to tell whether a particular DNA fragment comes from the bone or from someone who handled it.[4] To address this problem, several approaches have been developed. The first one was to establish as standard practice the extraction of aDNA under strict conditions in rooms that have been sterilized with ultraviolet radiation, bleach treatment of surfaces, and filtered air systems. These measures can help minimize the proportion of foreign DNA. Furthermore,

---

[3] Dabney, J., Meyer, M., and Pääbo, S. (2013). Ancient DNA damage. *Cold Spring Harbor Perspectives in Biology, 5*(7), a012567.

[4] Reich, D. (2018). *Who we are and how we got here: Ancient DNA and the new science of the human past.* Oxford: Oxford University Press, p. 31.

SEPARATING DNA FROM CULTURE   175

the molecules present at the moment of extraction are chemically tagged in order to distinguish them from molecules added during sequencing. Once DNA has been sequenced, several bioinformatic tools can be used to either remove contaminating sequences or estimate their proportion.[5]

The study of aDNA was instrumental in our understanding of recent human evolution. The fossil and aDNA evidence available support the conclusion that we have not been the only human species on Earth even in recent times, as other humans have coexisted and interacted with our ancestors about 30,000–40,000 years ago. These are the Neanderthals and the Denisovans, already mentioned in previous chapters. The sequencing of the Neanderthal genome showed that it was on average more similar to the genome of present-day humans from Eurasia than it was to those from Africa. At the same time, present-day humans from Eurasia have regions in their genome that are more similar to those in Neanderthals and more distant from other present-day humans. Overall, between 1% and 4% of the genomes of people in Eurasia have been derived from Neanderthals. Whereas this amount is small, it indicates that some kind of interbreeding took place between the Neanderthals and the ancestors of modern humans. The study of the DNA of the Denisovan individual also showed that the average difference of the Denisovan genome from present-day humans is similar to that of Neanderthals. Further analysis supported the conclusion that the Denisovan individual and the Neanderthals are both descended from a common ancestral population that was already separated from the ancestors of present-day humans. In other words, the Denisovans and the Neanderthals were closer to each other than they were to ourselves.[6]

But what can aDNA tell us that might affect our perception of who we are today? One might think that aDNA can point out our deep roots. Science and cultural studies scholar Venla Oikkonen analyzed the forms of belonging invoked by the analysis of DNA of the remains of two ancient humans: the "Cheddar man," a Mesolithic skeleton discovered in 1903 at Gough's Cave in Cheddar Gorge, Somerset, in Britain, and who was estimated to have lived

---

[5] Slatkin, M., and Racimo, F. (2016). Ancient DNA and human history. *Proceedings of the National Academy of Sciences, 113*(23), 6380–6387.

[6] Green, R. E., Krause, J., Briggs, A. W., Maricic, T., Stenzel, U., Kircher, M., et al. (2010). A draft sequence of the Neandertal genome. *Science, 328*(5979), 710–722; Reich, D., Green, R. E., Kircher, M., Krause, J., Patterson, N., Durand, E. Y., et al. (2010). Genetic history of an archaic hominin group from Denisova Cave in Siberia. *Nature, 468*(7327), 1053–1060. For popular accounts of the respective studies, see Pääbo, S. (2014). *Neanderthal man: In search of lost genomes*. New York: Basic Books; Reich, D. (2018). *Who we are and how we got here: Ancient DNA and the new science of the human past.* Oxford: Oxford University Press.

176 ANCESTRY REIMAGINED

around 10,000 years ago; and the "Kennewick Man," who was discovered in Washington State in the United States in 1996 and who was estimated to have lived around 9,000 years ago. The Cheddar man was included in discussions of *national belonging*, which concerned the origins of the inhabitants of Britain, given the historical record of invasions; *regional belonging*, which considered the people of Cheddar as the true descendants of the prehistoric populations of Britain; and *personal belonging*, for a particular person who was identified as the closest contemporary descendant of the Cheddar man. The Kennewick man also invoked different forms of belonging: *national belonging*, with reference to the origins of the inhabitants of the United States; *continental belonging*, which considered the evolutionary continuity of humans across all the American continent; and *ethnic/racialized belonging*, as the Kennewick man was considered by some as European American and for others as Native American. This indicates that the inferences that one can draw for belonging based on ancient DNA are variable and largely depend on context.[7]

But aDNA has also been used as a source of information for more recent events. In the remainder of this chapter, I focus on aDNA studies that concern the geographic area of modern Greece and its genetic history.

A Case Study: Ancient DNA and Greek Ancestry

Why consider the genetic history of Greece? There are two main reasons: (1) personal interest, as Greek ethnic identity is the one with which I identify and have personal experience, and (2) broader interest, because the related culture and language have existed for more than 3,000 years around the same area where modern Greece is found (Figure 9.1), and so all kinds of claims about continuity across time have been made. Therefore, it is interesting to consider what conclusions can be drawn from the respective aDNA studies. Whereas the study of modern DNA can provide information about the relatedness of the populations currently living in Greece, as well as between all those and neighboring populations in the Balkans, Turkey, and Italy, to the best of my knowledge there have been no studies aiming at describing the population structure within Greece as whole. What is available are studies

[7] Oikkonen V. (2018). *Population genetics and belonging: A cultural analysis of genetic ancestry.* Cham, Switzerland: Palgrave Macmillan, pp. 92–119.

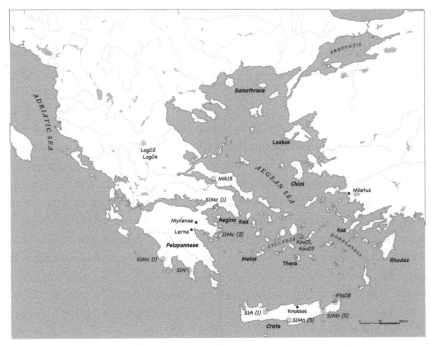

Figure 9.1 Map of Ancient Greece, indicating the main locations discussed in this chapter (adapted under the Creative Commons Attribution-Share Alike 4.0 International license). It also indicates the geographical locations of the samples of the ancient individuals in two studies discussed here (see Tables 9.1 and 9.2 for the details of the samples). Island names are in bold and italics (e.g., *Crete*); sample names are in italics (e.g., *Pta08*); city names are in regular fonts (e.g., Knossos); and island groups or seas are in capital letters (e.g., AEGEAN SEA).

of particular areas, which are of course of interest, such as Peloponnese or Crete.[8]

The idea of studying the genetic history of Greece was first suggested in 1993 by evolutionary geneticist Costas Krimbas, who was clearly inspired by the research program of Cavalli-Sforza and his colleagues, which we discussed in Chapter 8. At the time, only classical genetic markers (i.e.,

---

[8] Stamatoyannopoulos, G., Bose, A., Teodosiadis, A., Tsetsos, F., Plantinga, A., Psatha, N., et al. (2017). Genetics of the Peloponnesean populations and the theory of extinction of the medieval Peloponnesean Greeks. *European Journal of Human Genetics, 25*(5), 637–645; Drineas, P., Tsetsos, F., Plantinga, A., Lazaridis, I., Yannaki, E., Razou, A., et al. (2019). Genetic history of the population of Crete. *Annals of Human Genetics, 83*(6), 373–388.

178 ANCESTRY REIMAGINED

related to genes) were available, and what Krimbas argued for was the need to consider genetic, archaeological, demographic, linguistic, and other kinds of evidence, in order to reach conclusions about the major events that had determined the current population structure in Greece.[9] Twenty years later, in 2013, geneticist Costas Triantaphyllidis wrote a book in Greek titled *The Genetic History of Greece: The DNA of Greeks*, in which he put together all the available evidence at the time. His main conclusion in that book was that "All population genetic studies confirm, at least indirectly, the genetic continuity of Greeks. Modern Greeks are descendants of population groups that lived in this area of southeastern Europe since the Paleolithic era and have maintained the Greek language at least since the era of Myceneans until today."[10] Krimbas, and others, wrote critical reviews of that book.[11] A problem was that Triantaphyllidis had relied on various, not equally reliable, studies, as well different kinds of markers, to arrive at his conclusions. In 2018, Triantaphyllidis published a new edition of his book in English in which he attempted to reconstruct the genetic origins of Greeks. He concluded the book by writing: "From the DNA results of the Greeks and other Eurasian populations it can be concluded that the DNA signatures of the Greeks reflect the spread of the ancient Greeks during the prehistoric period in Eurasia and support the continuity of the Greeks' presence in their SE European corner."[12] If there is a genetic continuity of Greeks throughout time, this is something that should be evident in the study of aDNA. Let us then consider two recent studies that focused on individuals from civilizations of the Bronze Age in Greece, also called the Aegean Bronze Age, and the genetic relatedness between those and modern Greeks.

The Bronze Age is the period during which bronze replaced stone as the main material for tools and weapons. The Aegean Bronze Age is a period of approximately 2,000 years (around 3100–1100 BCE), and it concerns the history and the cultures in Crete, the Aegean islands, and the Greek mainland. It can be roughly divided into three periods: the Early Bronze Age (EBA) period (3100–2000 BCE); the Middle Bronze Age (MBA) period (2000–1600 BCE); and the Later Bronze Age (LBA) period (1600–1100 BCE). These cultures

---

[9] Κριμπάς Κ. (1993). Θραύσματα Κατόπτρου. Αθήνα: Θεμέλιο, pp. 123–167.

[10] Τριανταφυλλίδης, Κ. (2014). *Η Γενετική Ιστορία της Ελλάδας: Το DNA των Ελλήνων*. Θεσσαλονίκη: Εκδόσεις Κυριακίδη, pp. 341–342 (translation from Greek is my own).

[11] Κριμπάς, Κ. (2014). Σκέψεις για την καταγωγή των Ελλήνων. *The Athens Review of Books*, τχ. 51, Μάιος, σ. 27–29.

[12] Triantaphyllidis, C. (2018). *The genetic origins of the Greeks*. Thessaloniki: Kyriakidis Editions, p. 253.

SEPARATING DNA FROM CULTURE   179

have been studied by archaeologists during the last 130 years or so. Perhaps the best-known ones are the excavations in Knossos related to the Minoan civilization by Sir Arthur Evans, and in Mycenae related to the Mycenaean civilization by Heinrich Schliemann (see Figure 9.1). The area has generally been divided by scholars in three regions, which are labeled on the basis of the cultures of their inhabitants: "Minoan," after the legendary King Minos of Knossos for Crete; "Helladic," from the Greek word for Greece; "Hellas," for the Greek mainland; and "Cycladic," for the Cycladic islands or Cyclades.[13]

Based on the archaeological evidence about the possible interactions between Minoans, Mycenaeans, and the others, we can discern three main phases. In the first phase, around 1950–1750 BCE (Middle Bronze Age) pottery from Crete is found in various places in the Aegean islands such as Thera (known today as Santorini), Melos, Aigina, Kea, Rhodes, and even as north as Samothrace; it is also found in Lerna in mainland Greece, as well as in Miletus in Asia Minor (Figure 9.1). However, it seems that the level of contact between the different cultures is low. There also seems to be an import to Crete from the Cycladic islands, but it is limited. Overall, the direction of connection is mainly from Crete toward other places, but no kind of social changes seem to have taken place. However, the situation seems to change during the next phase, around 1750–1450 BCE (Middle to Late Bronze Age), when there seems to be a shift from "networks of exchange to networks of affiliation," with important social effects. For example, in Thera, various artifacts that could be characterized as Minoan are found; there are similarities to the Minoan prototypes, but there are also changes in their features and the techniques used to produce them, resulting in a kind of hybridization. The contact now extends to other places such as Lesbos, Chios, and Kos. Most importantly, the connections become richer and more complex; much of the evidence is about imports to Crete; however, there is also evidence for exchange between the Cycladic islands and the East Aegean. Finally, in the third phase, around 1450–1200 BCE (late Bronze Age), the Minoan influence continues despite the destructions that brought Minoan civilization to an end around 1450 BCE. There seems to be an overlap between Minoan and Mycenean influences, for instance in Miletus. However, later on the Mycenaean influence becomes stronger in Miletus, as well as in Rhodes, Kos, and the Cyclades. The Mycenaean influence thus seems to

---

[13] Shelmerdine, C. W. (2008). Background, sources, and methods. In C. W. Shelmerdine (Ed.), *The Cambridge companion to the Aegean Bronze Age*. Cambridge: Cambridge University Press.

180 ANCESTRY REIMAGINED

Table 9.1 Information about the 19 Ancient Individuals in the Lazaridis et al. Study

Symbol	Designation	No. of Individuals	Date (BCE)	Geographical Location (see Figure 9.1)
S1N	Neolithic	1	5400	Southern Peloponnese (Diros)
S1BA	Bronze Age	3	2800–1800	Southwestern Turkey
S1Mn	Minoans	10	2900–1700	Southern coast of central Crete (Moni Odigitria) Eastern Crete (Lasithi)
S1Mc	Mycenaeans	4	1700–1200	Western coast of the Peloponnese (Peristeria Tryfillia) Eastern Peloponnese (Galatas Apatheia) Salamis island
S1A	—	1	1370–1340	Western Crete (Armenoi)

have initially comprised the same network but also extended further than the Minoan to northern Greece and Cyprus, as well as Southern Italy, Sicilia, and Sardinia.[14]

With this historical and archaeological background, we can now consider the findings of the two aDNA studies. The first study, co-led by geneticist Iosif Lazaridis, was based on the analysis of DNA sequence data across the whole genome from what the researchers described as "genome-wide data from 19 ancient individuals, including Minoans from Crete, Mycenaeans from mainland Greece, and their eastern neighbours from south-western Anatolia." After DNA extraction from the ancient samples, the researchers aimed at capturing approximately 1.2 million single nucleotide polymorphisms (SNPs), but with a low sequencing depth (median 0.87X; see Box 1.2). As a result, it was not possible to reliably estimate the diploid genotypes (see Box 4.1). The DNA data from these 19 ancient individuals were combined with published data from 332 other ancient individuals. Table 9.1 summarizes the sample of the 19 ancient individuals that were the focus of this study.

---

[14] Knappett, C. (2017). Globalization, connectivities and networks: An archaeological perspective. In T. Hodos (Ed.), *The Routledge handbook of archaeology and globalization* (pp. 29–41). New York: Routledge.

SEPARATING DNA FROM CULTURE   181

Among the four research questions addressed in this study, I want to focus on two (questions 1 and 4 in the article):

[1] First, do the labels "Minoan" and "Mycenaean" correspond to genetically coherent populations or do they obscure a more complex structure of the peoples who inhabited Crete and mainland Greece at this time? . . .

[4] Fourth, how are the Minoans and Mycenaeans related to Modern Greeks, who inhabit the same area today?

Question 1 is of fundamental importance because, once answered, it forms a premise for answering the subsequent ones. If the answer in this question has problems, then there are problems for the answers to all the other questions because the groups that are actually compared are not the groups that the researchers had intended to compare. Question 4 is relevant to the topic of DNA ancestry tests and ethnicity estimates that I discuss in Chapter 10.

The researchers performed both a principal components analysis (PCA) and an ADMIXTURE analysis (see Chapter 7). In the PCA, the researchers computed the first two principal components based on an already available dataset of "1,029 present-day West Eurasians," which included already available data from 28 modern Greeks (from Greece and Cyprus) and two more from Crete, where they projected the data of 334 ancient individuals. The present-day populations formed two parallel clines from south to north in Europe and the Near East along PC2. The Minoans and Mycenaeans had a central position in the PCA, with ancient populations from mainland Europe and the Eurasian steppe to the left, ancient populations from the Caucasus and Western Asia to the right, and Early/Middle Neolithic farmers from Europe and Anatolia to the bottom, where the Neolithic samples from Greece were also found, thus being considered as distinct from the Minoans and Mycenaeans. The Bronze Age individuals from southwestern Turkey were also found to be distinct. ADMIXTURE analysis also showed that, when the clusters were eight or more, the Minoans and Mycenaeans had a component in common with several ancient individuals from Anatolia, as well as in Mesolithic/Neolithic samples from Iran and hunter-gatherers from the Caucasus in which it was the largest. That component was not found in the Neolithic samples of northwestern Anatolia, Greece, or the Early/Middle Neolithic populations of the rest of Europe. It only appeared in the

182 ANCESTRY REIMAGINED

populations of the Late Neolithic/Bronze Age in mainland Europe, which the authors attributed to migration from the Eurasian steppe.

Lazaridis and colleagues concluded that Minoans and Mycenaeans were genetically similar, with more than 75% of their ancestry derived from the first Neolithic farmers who entered Europe from western Anatolia and most of the rest from ancient populations in Caucasus and Iran. But they also differed in that the Mycenaeans had an additional ancestry, approximately 4%–16%, from a source related to the hunter gatherers of eastern Europe and Siberia. They concluded their article, by writing in the last paragraph: "The Minoans and Mycenaeans, sampled from different sites in Crete and mainland Greece, were homogeneous, supporting the *genetic coherency of these two groups*. Differences between them were modest, viewed against their broad overall similarity to each other and to the southwestern Anatolians, sharing in both the 'local' Anatolian Neolithic-like farmer ancestry and the 'eastern' Caucasus-related admixture" (emphasis added). Whereas they concluded the abstract of their article with following statement: "*Modern Greeks resemble the Mycenaeans*, but with some additional dilution of the Early Neolithic ancestry. Our results support *the idea of continuity* but not isolation in the history of populations of the Aegean, before and after the time of its earliest civilizations" (emphases added).[15] It is the genetic coherency of Minoans and Mycenaeans, the resemblance between the latter and modern Greeks, and the idea of continuity that I want to explore further.

The second study was co-led by computational biologist Anna-Sapfo Malaspinas and biological anthropologist Christina Papageorgopoulou. It was based on whole-genome data from six Bronze Age individuals, as shown in Table 9.2. The authors distinguished between the following different Bronze Age civilizations: the Minoan civilization in Crete (3200/3000–1100 BCE), the Helladic civilization in mainland Greece (3200/3000–1100 BCE) that also includes the Mycenaean during its last phase (1600–1100 BCE), and the Cycladic civilization in the Cycladic islands (3200/3000–1100 BCE). The analysis of the DNA of these six ancient individuals was more extensive in this study, as its authors noted, than the Lazaridis et al. study, because they also covered a set of 5,270,000 SNPs with an average sequencing depth of 3.73X.

---

[15] Lazaridis, I., Mittnik, A., Patterson, N., Mallick, S., Rohland, N., Pfrengle, S., et al. (2017). Genetic origins of the Minoans and Mycenaeans. *Nature, 548*(7666), 214–218.

SEPARATING DNA FROM CULTURE 183

Table 9.2 Information about the Six Ancient Individuals in the Clemente et al. Study

Symbol	Designation	No. of Individuals	Date (BCE)	Geographical Location (see Figure 9.1)
Mik15	Early Helladic	1	2890–2764	Manika, Euboea
Pta08	Early Minoan	1	2849–2621	Kephala Petras, Crete
Kou01	Early Cycladic	1	2464–2349	Ano Koufonisi, Cyclades
Kou03	Early Cycladic	1	2832–2578	Ano Koufonisi, Cyclades
Log02	Middle Helladic	1	1924–1831	Elati-Logkas
Log04	Middle Helladic	1	2007–1915	Elati-Logkas

Among the five research questions addressed in this study, I want to focus on two (questions 2 and 5 in the paper), which are more or less the same with the two from the other paper discussed above:

(2) What was the genetic affinity among the Helladic, Cycladic, and Minoan EBA civilizations (i.e., did their cultural differences entail population structure, and how did they relate to LBA populations such as the Mycenaeans)?

(5) How are Aegean individuals across the BA related to present-day Greeks who inhabit the same area?

Regarding question 2, the researchers wrote that when compared to other ancient Eurasian populations, the EBA Aegeans were similar to other Aegean BA and Anatolian populations, while being quite distinct from all Balkan populations. This was something they concluded both from multidimensional scaling analysis (which we can consider here as similar to PCA) and ADMIXTURE analysis. The authors wrote: "The genomic EBA homogeneity across cultures in the Aegean and parts of Anatolia may indicate that Aegean populations used the sea as a route to interact not only culturally but also genetically. This could have been the result of an intense network of communication in the Aegean, which has been well documented on the archaeological level." They also found from the ADMIXTURE analysis that the EBA Aegean populations had a main ancestry component (more than 65%) shared with Neolithic Aegeans, whereas the most of the

184 ANCESTRY REIMAGINED

remaining ancestry (17%–27%) came from Iran Neolithic/Caucasus re-lated to hunter-gatherers. This was supported by additional analyses. Another finding was that MBA individuals from northern Greece (Log02, Log04) were found to be quite distinct from the EBA Aegeans across all analyses. The primary feature distinguishing Log02, Log04 from the EBA populations was the higher proportion of European hunter-gatherer-like ancestry. Regarding the other research question about the relation be-tween Bronze Age individuals and modern Greeks, the researchers found that modern Greeks (individuals from northern Greece—Thessaloniki—and Crete) are closely related to the MBA individuals of northern Greece (Log02, Log04) in all analyses. They thus concluded that modern Greeks from northern Greece and Crete could be the descendants of Aegean EBA populations, with subsequent interbreeding with populations related to the Pontic-Caspian Steppe, something that was not the case for modern Cypriots included in the analysis.

The researchers thus arrived at some very interesting conclusions. "In sum-mary, these genomes from the Cycladic, Minoan, and Helladic (Mycenaean) BA civilizations suggest that these culturally different populations were *ge-netically homogeneous* across the Aegean and western Anatolia at the begin-ning of the BA" (emphasis added). In contrast, the MBA Aegean "population" differed, perhaps due to "additional Pontic-Caspian Steppe-related gene flow into the Aegean, for which evidence was seen in the newly sequenced MBA Logkas genomes." They also concluded that "Present-day Greeks—who also carry Steppe related ancestry—share 90% of their ancestry with MBA northern Aegeans, *suggesting continuity between the two time periods.* In contrast, *LBA Aegeans (Mycenaeans)* may carry either diluted Steppe-or Armenian-related ancestry (Lazaridis et al., 2017) [here they cite the other study we discussed in this section]. This *relative discontinuity* could be explained by the general decline of the Mycenaean civilization as previ-ously proposed in the archaeological literature" (emphases added).[16] Again, it is these claims about homogeneity and continuity that I am interested in exploring.

---

[16] Clemente, F., Unterländer, M., Dolgova, O., Amorim, C. E. G., Coroado-Santos, F., Neuenschwander, S. et al. (2021). The genomic history of the Aegean palatial civilizations. *Cell*, *184*(10), 2565–2586.

SEPARATING DNA FROM CULTURE   185

Do Modern Greeks Have Near-Mythical Origins?

"The Greeks really do have near-mythical origins, ancient DNA reveals." This is how the findings of the Lazaridis et al. study were described by science journalist Ann Gibbons in the news section of the prestigious journal *Science*. I must note at the outset that I am Greek, and I have absolutely no problem with the idea of continuity between ancient and modern Greeks, if it were to be well-established by the available data. I strongly believe that all humans are members of the same extended family, and that races, nations, ethnē, or any other social groups do not exist naturally; that is, there are no biological features that can be used to assign ourselves to one or other group. Nevertheless, if it were shown that the Greeks who today live in the area around modern Greece are the descendants of people who lived in the same area three or four thousand years ago, I would have no problem accepting it.

However, the studies discussed in the previous section arrived at conclusions about the continuity of Greeks across time on the basis of particular assumptions and conditions that seem problematic to me. It is the assumptions and the conditions, not the conclusion, that troubles me. If the studies had arrived at the conclusion that there is no continuity between Minoans/Mycenaeans and modern Greeks, I would be equally concerned. I am emphasizing this because there were different kinds of reactions to the Lazaridis et al. study in Greece in 2017 (the other study was published in 2021, just a few months before writing these lines), which depended strongly on political views. On the one hand, people on the far right cited the study to support their claims about Greek ethnic purity. On the other hand, people on the far left criticized the study exactly because it could be used to support such claims. But, again, I am not concerned with the conclusion reached, but with the assumptions and conditions underlying the studies. As I wrote, I would be equally concerned with these studies if they had arrived at the conclusion that there is no continuity between Minoans/Mycenaeans and modern Greeks. Let us then explore the assumptions and conditions of these studies.

First, one could be concerned about the sampling procedures. In their methods section, Lazaridis et al. wrote: "No statistical methods were used to predetermine sample size. The experiments were not randomized and the investigators were not blinded to allocation during experiments and outcome assessment." It is of course very important that the authors are clear about this. But this statement raises at least two questions. First, were the

186   ANCESTRY REIMAGINED

sizes of the samples used sufficient to ensure representativeness? If not, how can we be sure that the sample selection was not biased, in the sense that it does not represent all the available DNA variation? (see Chapter 8). Could ten "Minoans" and four "Mycenaeans" be a representative sample of the respective populations (assuming that such populations actually existed)? Furthermore, as the authors themselves noted in the methods, the DNA sequencing of the ancient samples had low coverage and so it was not possible to figure out their complete (diploid) genotypes. So they relied on these data to compute allele frequencies using a statistical method. Eventually, is it the case that what is presented in this article as data of Minoan and Mycenaean ancient DNA is a statistical estimate based on a rather partial, and perhaps biased, sample?

Second, how can we know that the researchers were not unconsciously biased in reaching their conclusions as they were not blinded during the study? I do not mean to suggest anything bad; but we know from psychological research how tempting it is to succumb to confirmation bias: a tendency to look for evidence that would confirm a hypothesis while failing to look for evidence that might disconfirm it (our tendency to infer causation from correlation is a good example).[17] This is human and natural, and it happens to all of us. But it is also a concern, in line with those that have been raised about similar biases in forensic investigations, which have led to proposals about temporarily "blinding" analysts to information they do not really need to have when they are analyzing and interpreting DNA tests, in order to reduce bias.[18]

In a similar manner, before reaching the conclusions about continuity and discontinuity between Bronze Age individuals and modern Greeks, Clemente and colleagues wrote: "Note that future work will be required to determine how representative the analyzed genomes of the Aegeans are of the BA Cycladic, Minoan, and Helladic cultures as a whole." I am honestly puzzled by how it is possible for them to write this—which is entirely accurate—and then make claims about continuity and discontinuity. In their paper, they argued that EBA individuals were distinct from LBA individuals, but is this something that their data really support? I think not, and indeed

[17] Nisbett, R. E. (2015). *Mindware: Tools for smart thinking*. London: Penguin Random House UK, p. 129.
[18] Thompson, W. C. (2013). Forensic DNA evidence: The myth of infallibility. In S. Krimsky and J. Gruber (Eds.), *Genetic explanations: Sense and nonsense* (pp. 227– 255). Cambridge, MA: Harvard University Press.

this is a concern also shared by experts working on ancient DNA. As Ludovic Orlando, director of research at the National Center of Scientific Research (CNRS) and aDNA expert wrote in a recent book: "Indeed, defining a group on the basis of an *a priori* limited number of individuals—one never sequences the entire fossil record—who themselves represent only a tiny fraction of the populations that have lived at a certain time, while rejecting the most distant individuals outside the group, only reinforces the initial definition of the group and gives the illusion of a people"[19] (emphasis in the original). What does this mean? The methods of analysis of genomic data may be solid, but we should be careful about what kinds of inferences we make from these data about history.

Another important concern is the conceptualization of Minoans and Mycenaeans as groups or populations that were "homogeneous" exhibited "coherency" with each other and "continuity" with modern Greeks. Lazaridis et al. very appropriately wrote in the supplementary information of their article, in a section titled "Note on Terminology":

> The terms "Minoan" and "Mycenaean" describe the Early-to-Late Bronze Age cultures of Crete and Late Bronze Age cultures of Greece respectively. We are aware that these terms were invented by early archaeologists of the 19th century and we do not wish to essentialize the complex past societies that they have been applied to or to view the world of the Aegean Bronze Age through these labels. We use them as the most common and accessible terms applied to these cultures, and also in order to empirically test their correspondence to genetically coherent clusters of the people of the Aegean Bronze Age.

This is very accurate, and I would have very much liked to have read it in the main article. However, even if the researchers did not wish to "essentialize" these groups, it is exactly what they did when they wrote about homogeneity, coherence, and continuity. As I explain in Chapter 2, homogeneity and fixity are two landmark features of essentialism. In fact, part of their research question 1 ("do the labels 'Minoan' and 'Mycenaean' correspond to genetically coherent populations?") is about homogeneity, and research question 4 ("how are the Minoans and Mycenaeans related to Modern Greeks, who

---

[19] Orlando, L. (2021). *L'ADN fossile, Une Machine à Remonter Le Temps: Les Tests ADN en Archéologie*. Paris: Odile Jacob, p. 211 (translation from French is my own).

188 ANCESTRY REIMAGINED

inhabit the same area today?") is about stability. Similar is the case for the other study that found that "these genomes from the Cycladic, Minoan, and Helladic (Mycenaean) BA civilizations suggest that these culturally different populations were genetically homogeneous" and that there was continuity between Middle Bronze Age northern Aegeans and modern Greeks.

Archaeologist Yannis Hamilakis was among the first to raise concerns about this issue, with reference to the Lazaridis et al. study. He pointed out that whereas research question [1] about whether the labels "Minoan" and "Mycenaean" correspond to genetically coherent populations seems open to investigation, the researchers had already affirmatively answered it in the way they asked it. By considering the categories "Minoan" and "Mycenaean" as distinct and by preassigning the skeletal samples to one of these categories, the researchers practically presupposed the correctness of their categorization—a logic he described as circular.[20] Hamilakis has also expressed a concern about the misuse of such studies: "we can see how the 19th-century discourse which emphasized the cultural-spiritual continuity of Hellenism, and its ability, as a superior culture, to absorb and 'civilize' others, despite the 'admixtures,' is being biologized today with the help of ancient DNA studies." In this view, these studies can help resurface and make explicit "the past, racial undertones of the previous national dogma."[21] In a recent essay, historian Joseph Maran also raised concerns about the circular logic of archaeologists assigning skeletal samples to a certain culture, and scientists accepting these assignments and using them to describe aDNA samples that thereafter accept the later as representative of that cultural only to reify the correctness of initial archaeological assignment. He also expressed concerns about the assumptions of the Lazaridis et al. study, by noting that it is problematic to use labels such as "Minoan" and "Mycenaean" as it implies that the respective societies in Crete and the Greek mainland were distinct for many centuries.[22]

This in turn raises the question: what is the meaning of designators such as "Helladic" or "Minoan" or "Mycenaean"? Let us consider the last one as

[20] Hamilakis, Y. (2017). Who are you calling Mycenaean? *London Review of Books blog.* https://www.lrb.co.uk/blog/2017/august/who-are-you-calling-mycenaean (accessed January 30, 2022).

[21] Greenberg, R., and Hamilakis, Y. (2022). *Archaeology, nation and race: Confronting the past, decolonizing the future in Greece and Israel.* Cambridge: Cambridge University Press, pp. 147–148.

[22] Maran, J. (2022). Archaeological cultures, fabricated ethnicities and DNA research. "Minoans" and "Mycenaeans" as case examples. In U. Davidovich, S. Matskevich, and N. Yahalom-Mack (Eds.), *Material, method, and meaning. Papers in Eastern Mediterranean archaeology in honor of Ilan Sharon.* Münster: Zaphon, 7–25.

SEPARATING DNA FROM CULTURE   189

example. The term "Mycenaean" was coined by Heinrich Schliemann to designate a Late Bronze Age society he discovered at Mycenae in Peloponnese, in 1876. Because of its influence on other parts of Greece, as we have seen, the term is also used to refer to societies in central and southern Greece as well as the Aegean Sea, Crete, and Asia Minor that came to share similar cultural traits, a common Mycenaean culture, during the Late Bronze period. These include similarities in architecture, art styles, and technologies, and political, economic, and administrative structures. However, the commonalities were also accompanied by regional variations in styles and practices, and so the term "Mycenaean" does not reflect homogeneity. Interestingly, these commonalities appear to be features of the social elites, most often associated with the palaces, rather than the respective societies as a whole. So overall, it seems that the term "Mycenaean" describes a recognizable cultural framework with regional variation, which was adopted deliberately. This common cultural framework seems to have been an indicator of a common identity: the common cultural traits (styles, practices, rituals, language, etc.) could serve to distinguish a particular social group from others.[23] What does this entail? That Mycenaeans were not a homogeneous social group with genealogical connections, but a broad, consciously and socially constructed one that covered a large part of Greece. As historian Guy D. Middleton put it: "In conclusion, 'the Mycenaeans' exist only in one reality—that of modern scholarship. They have been constructed from a complex intellectual background that draws upon archaeology, mythology, and modern cultural assumptions and values, as well as upon the psychology of individuals."[24] Is it then possible for its genetic variation to be represented by the few ancient individuals in two studies considered here?

In fact, these two aDNA studies did not even provide evidence that these ancient individuals actually lived in the areas in which the specimens were found or if they just happened to die there. This is something that might be established with a strontium isotope analysis of the teeth of these individuals. The strontium isotopes have different masses ($^{87}Sr/^{86}Sr$) and are found in

[23] Mac Sweeney, N. (2008). The meaning of "Mycenaean." In O. Menozzi, M. L. di Marzio, and D. Fossataro (Eds.), *SOMA 2005: Proceedings of the IX Symposium on Mediterranean Archaeology, Chieti (Italy)*, February 24–26, 2005 (pp. 105–110). BAR International Series 1739. Oxford: British Archaeological Reports.

[24] Middleton, G. D. (1995). Mycenaeans, Greeks, archaeology and myth: Identity and the uses of evidence in the archaeology of Late Bronze Age Greece. *Eras Journal, 3.* https://www.monash.edu/arts/philosophical-historical-international-studies/eras/past-editions/edition-three-2002-june/mycenaeans-greeks-archaeology-and-myth-identity-and-the-uses-of-evidence-in-the-archaeology-of-late-bronze-age-greece

190 ANCESTRY REIMAGINED

different relative concentrations in different geographical areas. Given that ancient individuals usually gathered their food from the place in which they lived, and not from other places far away, we can figure out the relative concentrations of the strontium isotopes in their skeletons, absorbed through diet, and from this infer where these people grew up. In fact, because certain parts such as the enamel of the molars that develop during childhood also absorb strontium, it is possible to find whether a person lived their life in a single place or moved around.[25] Such information was available for only one sample in the Clemente et al. study. At the same time, even an analysis of strontium isotope ratios might not reflect the place of origin of those ancient individuals exactly because it is actually indicative of nutrition. If people indeed used local products for their nutrition, we can make inferences about where they lived. But food was exchanged among peoples since ancient times, and so it is possible that the strontium ratio would reflect the origin of the food rather than the origin of the people. For instance, if we performed such an analysis with the remains of Roman soldiers, what we would find would likely be the place of origin of their food supply, not the place of origin of the soldiers themselves.[26]

But if one were to consider the findings of this kind of studies, some very interesting conclusions might arise. One study aimed to figure out whether the human remains found by Schliemann in Mycenae belonged to people who had always lived there or to people who had come from elsewhere. In particular, strontium isotope analysis was applied to samples of dental enamel from eleven adult individuals. What was found was that only two among the eleven individuals could be characterized as locals, because the others had higher than expected strontium ratios. The author discussed all the limitations of the methods used and pointed out that the results should be considered as tentative, arguing that the higher strontium ratios could reflect just as well not only the nonlocal origin of the nine individuals but also the consumption of food from the periphery of Mycenae (up to 5 to 10 km away) where the local strontium isotope ratios are higher. But if one accepted that the results indicated a nonlocal origin, there were some interesting conclusions to make. The only two individuals who were identified as female had been shown to be nonlocal. According to the author, this finding

---

[25] Krause, J., and Trappe, T. (2021). *A short history of humanity: How migration made us who we are.* London: WH Allen, pp. 142–143.

[26] Brather, S. (2016). New questions instead of old answers: Archaeological expectations of aDNA analysis. *Medieval Worlds, 4,* 22–41.

SEPARATING DNA FROM CULTURE   191

might reflect marital patterns and the nonlocal origin of females associated with Mycenae elites.[27] Historian Jonathan Hall has suggested that this study presents an interesting theoretical proposition: "what if the ruling elites at Mycenae regularly practiced exogamy with Minoan elites? Recurrent inter-marriage over several generations might account for the genetic results that Lazaridis and colleagues found but the *obligation* to practice exogamy would simultaneously reassert the *differences* between the two societies/cultures—which would be a clear instance of where ethnicity and genetics don't match up" (Jonathan Hall, personal communication).

A last point to note is that the claims about continuity between Bronze Age civilizations and modern Greeks cannot be really established based on the resemblance of the aDNA samples from skeletons found in Greece and the DNA from people living today in the same area. As Maran put it: "Due to the myriad patterns of mobility in the last 8,000 years, today's populations of Mediterranean descent are a mixture of essentially iden-tical genetic components, mostly of Near Eastern derivation. They are thus so interrelated that it is totally naive to expect a close and exclusive corre-spondence between a modern population and a specific group of the past."[28] This last point is exactly what geneticists Coop and Ralph established in their 2013 study, discussed in Chapter 4. They showed that every person who has lived 1,000 years ago and has had descendants would be an ancestor of every European living today. If we cannot establish a distinct continuity between people who lived 1,000 years ago in a particular place in Europe with people who live there today, how can we do this between people who lived more than 3,000 years apart? Of course, this does not deny the fact that similarities at the level of DNA can be found between individuals who lived in the same area thousands of years apart. But this does not entail that the respective populations to which they belonged have had continuity across all this time.

Perhaps more modest titles in these two aDNA papers would have allowed for more modest claims. For instance, they might have simply referred to the geographical area where particular individuals were found rather than make claims about "Minoans and Mycenaeans" or "Aegean palatial civilizations."

---

[27] Nafplioti, A. (2009). Mycenae revisited Part 2. Exploring the local versus non-local geographical origin of the individuals from grave circle A: Evidence from strontium isotope ratio (87Sr/86Sr) anal-ysis. *Annual of the British School at Athens, 104*, 279–291.

[28] Maran, J. (In press). Archaeological cultures, fabricated ethnicities and DNA research. "Minoans" and "Mycenaeans" as case examples. In U. Davidovich, S. Matskevich, and N. Yahalom-Mack (Eds.), *Material, method, and meaning. Papers in Eastern Mediterranean Archaeology in Honor of Ilan Sharon* (p. 14). Münster: Zaphon.

192 ANCESTRY REIMAGINED

Jonathan Hall has clearly shown that ethnic identity in Greek antiquity was a contingent phenomenon depending on particular contexts. As he noted: "Ethnic identity is not a 'natural' fact of life; it is something that needs to be actively proclaimed, reclaimed and disclaimed through discursive channels. . . . To understand the ethnic group, we must learn how the ethnic group understands itself."[29] Because of these and the issues discussed in this section, I would add that there is nothing that we can figure out about ethnic identity from DNA.

How We Can Best Study Human Ancestry with aDNA

There are some fundamental questions about aDNA research. For instance, one might question the validity of inferring the existence of Denisovans from aDNA and think of them as a kind of reification. The Denisovans are not even accessible for study to paleontologists, because we have only a few, small bones and so we cannot infer if they had any distinctive anatomical features. Any new specimen that might be a Denisovan could only be confirmed as one by paleogeneticists—this is how geneticists studying aDNA are called. A response to this concern could be that even though Denisovans are diagnosable at present only by their DNA, this doesn't make them any less "real" as a biological entity. Genomic characteristics (e.g., via genetic barcoding) are currently used to distinguish many living species, and the differences in Denisovan DNA sequences are sufficient for classification. In the future, we might come to have more information and better knowledge, but for now such different perspectives may exist.

There have also been different kinds of reactions from archaeologists toward aDNA studies. Some archaeologists have reconceptualized their work and collaborate with paleogeneticists. Others are skeptical about aDNA studies. But what is clear is that the different disciplines must find common ground and collaborate effectively if we want to better understand our past. And, indeed, there already exist collaborations among people from different disciplines toward producing a richer and more complex picture of the human past.[30] For instance, perhaps as a first step, scholars from various

---

[29] Hall, J. (1997). *Ethnic identity in Greek antiquity*. Cambridge: Cambridge University Press, pp. 182, 185.

[30] Callaway, E. (2018). The battle for common ground. *Nature*, 555(7698), 573–576.

SEPARATING DNA FROM CULTURE   193

disciplines have come together to discuss and clarify the nomenclature they use.[31]

An interesting approach suggested by medieval historian Patrick Geary and geneticist Krishna Veeramah is to move from geographically and temporally broad "top-down" paleogenomic studies, such as the ones we have considered in this chapter, to more "bottom-up" studies focused on particular communities. In such a study, they analyzed aDNA from skeletons from 63 graves in two cemeteries, Szólád in Hungary and Collegno in Northern Italy. These cemeteries had been associated in historical texts with the Longobards, who ruled a large part of Italy for more than 200 years after their invasion in 568 CE. The analysis revealed that each cemetery was primarily organized around one large male-dominated family tree. Both of these had ancestry that was not found typically in those regions. In addition, strontium data showed that the first-generation individuals in Collegno were probably not local, in contrast to the second- and third-generation individuals. This was considered consistent with the proposed long-distance migration to Northern Italy.[32] So the suggestion here is that rather than analyzing a few samples dispersed in a large area and draw on those data to arrive at generalizations, it might be better to study as many samples as possible from particular narrow areas in order to get a more detailed view of DNA variation at a specific place and time.

All this is important to keep in mind because companies already exist that suggest that ethnicity estimates could be more reliably produced via the use of aDNA from the actual individuals of past populations and offer such comparisons. For instance, a company called "My True Ancestry" invites people to upload their raw DNA data that they have obtained from DNA testing companies and "in just 10 minutes" they "will connect you to your ancient past." Their mission statement is this:

> Our mission is to help you take your DNA results a step further. Discover your ancient relatives by comparing yourself to thousands of ancient samples from real archaeological sites. Our simple and secure service puts

---

[31] Eisenmann, S., Bánffy, E., van Dommelen, P., Hofmann, K. P., Maran, J., Lazaridis, I., et al. (2018). Reconciling material cultures in archaeology with genetic data: The nomenclature of clusters emerging from archaeogenomic analysis. *Scientific Reports, 8*(1), 1–12.

[32] Amorim, C. E. G., Vai, S., Posth, C., Modi, A., Koncz, I., Hakenbeck, S., et al. (2018). Understanding 6th-century barbarian social organization and migration through paleogenomics. *Nature Communications, 9*(1), 1–11; Veeramah, K. R. (2018). The importance of fine-scale studies for integrating paleogenomics and archaeology. *Current Opinion in Genetics & Development, 53*, 83–89.

194 ANCESTRY REIMAGINED

10000 years of history, and over 85 ancient civilizations from around the world, at your fingertips.[33]

Perhaps not surprisingly, Myceneans are one of these ancient civilizations.

Even if aDNA can be used to produce a more direct picture of past populations compared to modern DNA, because with the former we can "read" the DNA sequences of past individuals rather than infer them from the latter, this is not like a direct observation of the past. Rather, aDNA, like modern DNA, is analyzed with laboratory methods and complex mathematical analyses that are necessary for comparisons and conclusions. Most important, ethnic identities such as Minoans, Mycenaeans, Celts, Longobards, Vikings, or whatever are the products of culture, not of biology. Only if we culturally establish that such civilizations have existed, can we make inferences about their cultural features, ways of life, and so on. DNA has nothing to say about them, because there is no distinct Minoan, Mycenaean, Celtic, Longobard, or Viking DNA. We can be proud of our ethnic identity and our culture, but DNA has nothing to do with this. I am Greek because of my upbringing, because of the origin of my ancestors, and I do not need my DNA to show anything about it. Even if it did show any other origin, it would not change who I am. And if I am proud to suggest to my non-Greek friends to visit Knossos, Mycenae, and any other of the numerous archaeological sites of interest in Greece, it is because these are found in the country I was born in, and because this is the culture in which I was brought up. I do not need to establish any continuity with ancient Greeks in DNA to be proud of this history.

We have now reached the point of having all the required background for understanding the results of DNA ancestry testing.

---

[33] https://mytrueancestry.com/en

10

Finding Meaning in Our Ancestry Testing

The Implications of the Paradoxes about Ancestry for DNA Ancestry Testing

DNA ancestry testing is based, and depends, on the methods and the findings of human evolutionary genetics, discussed in Chapters 6–9 of the present book. Only if we consider and understand these can we make sense of what DNA ancestry tests aim at doing and what they can really do. Our foray in the science of human evolutionary genetics in the previous chapters has revealed particular paradoxes, which are crucial for understanding DNA ancestry testing. Let me summarize them here:

1. Ancestry is collective and global, but we think of it primarily as individual and local (Chapter 6).
2. Ancestry is mostly about relatedness, but we privilege distinctiveness (Chapter 7).
3. Ancestry groupings are human inventions, but we consider them as natural (Chapter 8).
4. Ancestry depends on culture, but we think of it as based on DNA only (Chapter 9).

These paradoxes have some direct implications for ancestry testing, which are the following:

1. The aim of DNA ancestry testing is perceived as clear, even though ancestry is a multidimensional concept.
2. We are both related and distinct, but in DNA ancestry testing discourse we privilege distinctiveness.
3. In DNA ancestry testing, DNA data from individuals and groups are used to assign ancestries to each other, resembling a chicken-egg situation.

Ancestry Reimagined. Kostas Kampourakis, Oxford University Press. © Oxford University Press 2023.
DOI: 10.1093/oso/9780197656341.003.0010

4. To assign humans to ancestry groups in DNA ancestry testing, DNA is considered as an objective criterion and culture a subjective one, even though the categorizations related to the former are often derived from the latter.

Let us consider these implications one by one.

The first implication pertains to our understandings of the concept of ancestry. The reason I call it a multidimensional concept is that it can be perceived differently in at least three dimensions:

- *Time*: Ancestry can be ultimate, with reference to hundreds or thousands of years ago (mtDNA and YDNA), as well as proximate with reference to few generations back (autosomal DNA).
- *Space*: Ancestry can be local, with reference to geographical regions within countries, but also global with reference to the origin of our species in Africa.
- *Level*: Ancestry can be personal with reference to the individual level but also collective with reference to the population level.

Therefore, any discourse about ancestry has the potential to be misleading if the above dimensions are not made clear. What are we interested in? What can we really know about? DNA ancestry testing can provide information about all these dimensions. Therefore, whenever we talk about DNA ancestry testing, and ancestry more broadly, it may be useful to specify what we are interested in and kind of answers we seek.

The second implication is that whereas the evidence from human evolutionary and population genetics indicates that all humans share around 99.9% of our DNA due to our common descent and that therefore we are all related, at the same time for pragmatic purposes (e.g., in order to study human DNA variation) scientists divide humans to distinct groups. As a result, even though the scientific evidence points to a panhuman belonging because we share most of our DNA and because the ancestry of all of us ultimately goes back to Africa, we end up prioritizing regional belongings that stem from the relatively tiny differences (around 0.1%) in our DNA. In other words, we unconsciously blind ourselves to how much we have in common and focus on the differences among ourselves that are millions in absolute numbers but still relatively few if we consider the billions of identical DNA bases that we share. This is like going to a forest and try to distinguish among

FINDING MEANING IN OUR ANCESTRY TESTING   197

the various trees that exist there, thereby highlighting the differences among them and forgetting that they are all trees. Distinguishing among the various types of trees is a scientifically legitimate action for the purposes of studying the respective variation—and this is exactly what scientists do with human DNA variation. This is not a problem at all. The problem lies at the inferences we might make from the results of DNA ancestry testing about how humanity can be "naturally" divided into distinct groups. Perceiving these groups as absolute, and dividing ourselves to "us" and "them" groups is a practice that finds no support in the study of human DNA diversity.

The third implication stems from the information that DNA ancestry tests can provide about the human past. This information can be both about individual and collective pasts. But we should not forget that these individual and collective pasts are mutually related. The companies analyze the DNA of individuals who take their tests and then they accumulate their data to form datasets about groups, which they designate in an abstract manner—usually informed by ethnicity or geography. Then they compare data from new individuals to these datasets and assign these new individuals to one of these groups. In this way, individual pasts can be inferred from collective pasts that have been formed based on other individual pasts, perhaps resulting in an awkward tautological relationship, which resembles a chicken-egg situation: individuals are used to define groups that are in turn used to define individuals, who can in turn stand as representatives of those groups. The problem here is that there is no independent, "objective" point of reference for defining groups or individuals: they just define each other.

The fourth and final implication relates to the designation of the aforementioned groups. As mentioned, features such as ethnicity or geography are used to divide people into groups. Once DNA markers are found that can be used to distinguish between these groups, these markers are thereafter perceived as the distinctive features of the respective groups (this is what AIMs are supposed to do), while also being considered as more objective than the cultural features used to make the initial divisions (this what the ethnicity estimates or the ancestry composition that the companies offer are about). In other words, biological markers are considered as more objective than non-biological markers, even though the former were selected thanks to divisions initially made on the basis of the latter. To give an example, I and many others are Greeks because all my grandparents were born in Greece. If we participate in a study aiming at choosing distinctive "Greek" DNA markers, these could then be taken to be representative of Greeks everywhere in the world.

198 ANCESTRY REIMAGINED

Once this was done, they could be used about deciding whether other people are Greek, perhaps independently of where they and their grandparents were born. DNA could thus overthrow the place of known origin as a criterion, although it was the latter that made the former important in the first place.

With these implications considered, let us now see what the ethnicity estimates of DNA ancestry testing can reveal.

Ethnicity Estimates and Their Limitations

All ancestry companies explain in detail how they analyze DNA samples to produce ethnicity estimates. However, a detailed comparison of these methods falls outside of the scope of the present book, especially as excellent books on this topic already exist.[1] Therefore, in this section I focus on the methods of Ancestry because it has the largest number of customers and because they have recently published a "white paper," the "Ethnicity Estimate 2021 White Paper" of Ancestry,[2] in which they explain their methods very transparently and very clearly. However, they provide no definition of ethnicity in order for customers to clearly understand what the company means by this term. The ethnicity estimate is described as "a fast, sophisticated, and accurate method for estimating the historical origins of customers' DNA going back several hundred to over 1,000 years." From this phrase one might infer that ethnicity has to do with one's historical origins, but this is all that people get by way of definition.

The procedure begins with customers receiving a kit that they use to send their saliva sample to the company. After extracting DNA from the saliva sample, which is done with a chemical process that separates DNA from proteins and other molecules, it is possible to use DNA sequencing in order to analyze the customer's DNA for 300,000 single nucleotide polymorphisms (SNPs). They explain further that they do not look at each SNP in isolation but at several SNPs together as a haplotype (see Chapter 6), as well as that each person's DNA is divided into 1,001 smaller segments that are called "windows." Then in order to make an ethnicity estimate, they compare "a

---

[1] A very informative and readable one is Bettinger, B. T. (2019). *The family tree guide to DNA testing and genetic genealogy*. Cincinnati, OH: Family Tree Books. Also recommended are Krimsky, S. (2021). *Understanding DNA ancestry*. Cambridge: Cambridge University Press; McHughen, A. (2020). *DNA demystified: Unravelling the double helix*. Oxford: Oxford University Press.

[2] https://www.ancestrycdn.com/support/us/2021/09/ethnicity2021whitepaper.pdf (accessed January 31, 2022). All quotations in this section come from this white paper, unless otherwise noted.

FINDING MEANING IN OUR ANCESTRY TESTING 199

customer's DNA to the DNA of people with long family histories in a particular region or group, what we call the reference panel, and looking for segments of DNA that are most similar." The ethnicity estimates result from an algorithm that compares the DNA sequence of a customer to those of other customers already in the reference groups and calculates the similarities (matches) and the differences. The company has gathered together DNA markers from people who are considered to represent the populations in particular regions of the world, because they and their ancestors of several generations have lived in those regions (similar to the assumptions of the research studies that we considered in Chapters 7 and 8). These people then are considered to form reference groups, one for each particular region. Currently, Ancestry uses 77 reference groups that include data from more than 55,000 people.[3]

The company is explicit about the limitations that this procedure has, as well as about the fact that their reference groups of people who live today are just proxies for ancestry. "Identifying the best candidates for the reference panel is key to providing the most accurate ethnicity estimate possible from a customer's DNA sample. Under perfect circumstances, we would construct our reference panel using DNA samples from people who lived hundreds of years ago. Unfortunately, it is not yet possible to reliably sample historical populations in this way. Instead, we must rely on DNA samples collected from people alive today and focus on those who can trace their ancestry to a single geographic location or population group." But as I have already explained earlier, this is based on the assumption that these individuals can indeed trace their ancestry back in time for several generations, which is not always possible. As I have noted, I would qualify to be included in the reference group "Greece," as I know that all my great-grandparents (most of whom were born in the late 19th century, and three of whom I actually met) have always lived in Greece. But I do not know much about what happened before them and where their own great-grandparents lived.

Indeed, the Ancestry white paper states: "When asked to trace familial origins, most people can only reliably go back one to five generations, making it difficult to find individuals with knowledge about more distant ancestry. This is because as we go back in time, historical records become sparse, and the number of ancestors we have to follow doubles with each generation."

---

[3] https://www.ancestry.com/dna/lp/ancestry-dna-ethnicity-estimate-update (accessed January 31, 2022).

200 ANCESTRY REIMAGINED

We have discussed all this in Chapter 4. And they continue: "Fortunately, knowing where someone's recent ancestors were born is often a sufficient proxy for much deeper ancestry. In the recent past, it was much more difficult and thus less common for people to migrate large distances. Because of this, the birthplace of a person's recent ancestors often represents the location of that person's deeper ancestral DNA." Well, as I have shown, these assumptions, also made by Cavalli-Sforza and others, may not always be valid. Knowing where a person and his recent ancestors were born and lived may be a good proxy, but it is not necessarily a sufficient condition for inferring ancestry because some ancestors a little further back in time may have moved to that place from elsewhere, and it just happens that nobody knows about this today. Limited migration entails that the members of a particular group were only mating with members of their own group, and not with those of others. But no actual population really meets these criteria. Beyond these, in order to form a reference group, Ancestry also exclude people that are found to be closely related. The criterion for this is to have segments that are identical by descent (IBD) that are longer than 20 cM (see Chapter 4). They also exclude samples when the DNA data about ethnicity are not in agreement with the information that a person has reported about it. Finally, they perform a principal components analysis (Chapter 7) to identify people who form clusters, and these are eventually the ones that are included in the reference group for each region after several refinement procedures.

The basic premise behind the ethnicity estimate is that "Two chromosomes from the same geographic region or the same population will share more DNA with one another than will two chromosomes from different regions or groups. So two pieces of DNA with a historical connection to Portugal will have more DNA in common than will a piece of DNA from Korea and a piece of DNA from Portugal." This is where you have to be very careful in order to understand what is going on. "If, for example, a section of a customer's DNA looks most similar to DNA in the reference panel from people from Norway, that section of the customer's DNA is said to be from Norway, and so on. The end result is a portrait of a customer's DNA made up of percentages of the 77 regions contained in the reference panel." What this means is that they do not find any DNA markers that are distinctive of particular regions or countries, such as Norway. What they find is that the DNA sequence of a person is more similar to those of people in one region, such as Norway, rather than another, such as Finland. This means that the particular section of DNA of that person

FINDING MEANING IN OUR ANCESTRY TESTING   201

is labeled as "Norway," even though it also has similarities to sequences from Finland—they are just fewer. In other words, whereas the comparison actually indicates that a particular portion of DNA is "more similar to Norway than Finland," the results only mention Norway and omit Finland.

The company is clear about this:

> Because the probability of a specific pair of alleles appearing at a certain position in the DNA varies for each of our 77 regions, we can use that information to tell us which region a stretch of DNA most likely came from. For example, if AA at a particular position is more common in people from Spain, someone with AA at that location might have a higher chance of having Spanish ancestry. It is important to keep in mind that an AA at this particular position just makes it more likely the DNA comes from Spain. Plenty of people from Portugal, France, or even Korea might have AA at this position as well. The ethnicity estimate uses the probability at all positions within a window to determine where the DNA most likely came from.

So this means that Jay's results that we saw in the Preface do not mean that he is 55% Irish, 30% British, 5% French-German, 3% Spanish-Portuguese, 3% Italian-Greek, 3% Scandinavian, and 1% Turkish. They rather mean that 55% of Jay's DNA is more similar to the DNA of the company's reference group from Ireland than to the company's reference group from other countries, 30% of Jay's DNA is more similar to the DNA of the company's reference group from Great Britain than to the company's reference group from other countries, and so on. In short, *any ethnicity estimate is relative, not absolute.*

There are two important consequences of this relativity. The first is that the estimates do not necessarily agree with the family history of a person. Migration may take people away from where their ancestors had lived, and this might be reflected in their DNA—in the same way it might be reflected in their surname. For instance, George Stephanopoulos, former White House communications director, was born in the United States, but his surname clearly indicates origin from Greece (but I have met people with Greek surnames who did not know if they had any origin in Greece). This is why people may be surprised by their results. Of course, to this we must add what we discussed in detail in Chapter 4: that only few of a person's genealogical ancestors are also that person's genetic ancestors. This simply means that each one of us has inherited their DNA only from very few of our ancestors. So some people are not surprised if they know the origins of their ancestors;

some may be surprised if they do not; and some, like Dani Shapiro, may find out that their ancestors were not really the ones they thought.

The second important point, which may also lead to a disagreement with what one knows about their family history, is that this relativity is the reason that a person's ethnicity estimate might change across time. Ancestry explained: "We keep adding to the reference panel and optimizing the algorithm we use to compare your DNA to our reference groups, allowing us to add more regions and leading to improved precision in your results over time. DNA science is always evolving, so remember to look at your latest AncestryDNA® results whenever we make an update. As more people take the AncestryDNA® test and science progresses, we can give you a higher level of precision on your results." As more people take tests and are thus added to the respective reference groups across time, the referent to which a person's DNA is compared changes. This is why regions that were not initially well represented in the reference groups, may later appear in a person's results. As time goes by, the reference groups become larger, and so the geographical resolution increases. What initially could have been a reference group Italy-Greece may later be divided into smaller groups that correspond to smaller regions within Italy or within Greece. Indeed, one of the regions added in September 2021was the Aegean Islands (Figure 9.1). This is why the company asks customers to look for updates.[4]

It is useful to point out here that even though the results are described as "ethnicity estimates," the reference groups are not labeled with reference to ethnicities, but rather with reference to geographical regions. For instance, there is no category "Greek" but "Greece & Albania." This is further divided into two broader regions: "Albania, Northern Greece & North Macedonia" that comprises four regions (Albania and Western North Macedonia; Northern Greece; Southern Albania and Northern Greece; West Central North Macedonia), and "South Greece" that comprises seven regions (Central Greece; Ionian Islands; Peloponnese; South Peloponnese; Southeast Peloponnese; Southwest Peloponnese; West Peloponnese). Finally, the recently added group "Aegean Islands" includes Crete, Dodecanese Islands, East Central Aegean Islands, North Aegean Islands, and West Central Aegean Islands.[5] The criteria for this classification are not given, and it is not clear how one is supposed to make sense of them. On the one hand,

[4] https://www.youtube.com/watch?v=eV4-aSsMEg (accessed January 31, 2022).
[5] https://support.ancestry.com/s/article/List-of-AncestryDNA-Regions?language=enUS (accessed January 31, 2022).

FINDING MEANING IN OUR ANCESTRY TESTING   203

Albania and Greece are two distinct countries with different languages, religion, traditions, and more. On the other hand, during the last 30 years many Albanians have been living in Greece, and in the most recent generations one can find many Greek citizens of Albanian origin. All I am trying to say is that these labels make sense, if they do, only in geographical terms and not ethnic or other such terms. However, ethnic terms are used and the final outcome is described as "ethnicity estimate," whereas it might make more sense to be described as "geographical origin estimate." Interestingly, the Aegean Islands that are part of Greece are presented as a reference group distinct from the Greece/Albania one. Most interestingly, the reference group for the Aegean Islands consists of 366 samples, whereas the one for Greece and Albania consists of 482 samples; but the Aegean Islands include less than 10% of the total population of Greece. Why? Ancestry does not seem to provide any justification for this.

A last point to note is that ethnicity estimates are statistical estimates within a range of probabilities rather than absolute numbers. As Ancestry states: "Ethnicity estimates are not an exact science. The percentage AncestryDNA reports to a customer is the most likely percentage within a range of percentages. . . . So, for example, we might report someone as 40% England & Northwestern Europe with a confidence range of 30–60%. This means that they are most likely 40% England & Northwestern Europe but they could be anywhere between 30% and 60% England & Northwestern Europe." My parents and grandparents come from different parts of Greece. My maternal grandfather came from Southeast Peloponnese, whereas my maternal grandmother came from Samos, an island in the Eastern Aegean; both my paternal grandparents came from Crete. So, according to the regions that Ancestry uses, in theory my ancestry would be 25% "South Greece" and 75% "Aegean Islands." But is it that easy to figure this out? No, not at all, and again Ancestry is clear about the uncertainty that this brings in:

> What this means is that if a customer's ethnicity estimate has many nearby regions, their ranges will most likely be larger than if it contained more distant regions. For example, while we may be fairly certain that a customer has 50% Korea and 50% Portugal ancestry (and therefore small ranges), we may be less sure about a customer who gets 50% Spain and 50% Portugal. It is relatively easy to tell Korea from Portugal but relatively hard to tell Portugal from Spain. This may be reflected in the larger ranges for the second customer. But it is important to keep in mind that we are very

204 ANCESTRY REIMAGINED

confident of the European heritage of customer two, we are just less certain about how much ancestry is derived from Portugal and how much from Spain. It is worth noting that, in general, as we increase the precision of our regions (e.g., breaking Ireland & Scotland into two separate regions), the ranges may become larger, and that this is due to the fact that DNA from neighboring regions is still very similar.[6]

To summarize, there are important limitations and assumptions in the ethnicity estimates, and Ancestry is clear and transparent about these. But it is also important for test-takers to understand all them well, in order to make sense of their results.

How Ought We to Interpret the Results of DNA Ancestry Testing?

With caution!

There are at least five important points to keep in mind while reading the DNA ancestry test results:

1. *Your DNA ancestry results are not really about your ancestry.* What ancestry tests provide are just probabilistic estimates of similarities between the test-takers and particular reference groups, which consist of people who live today. But being related genetically to people who live today somewhere does not necessarily mean that your ancestors came from that place. The tests point to people with whom you have common ancestors, but not to the ancestors themselves. As population geneticist Graham Coop cogently put it, saying that "Graham is genetically similar to the GBR 1000 genome samples (on the first 10 principal components)" is a more accurate description of the findings of DNA analyses than "Graham has Northwestern European genetic ancestry." As Coop noted, "The former sounds a little more awkward, but that awkwardness reflects the truth of how these labels work and comes with many fewer built-in assumptions and pitfalls."[7] Furthermore, as

---

[6] https://www.ancestrycdn.com/support/us/2021/09/ethnicity2021whitepaper.pdf (accessed January 31, 2022).

[7] https://gcbias.files.wordpress.com/2022/07/genetic_similarity_and_genetic_ancestry_groups_current.pdf (accessed July 14, 2022).

FINDING MEANING IN OUR ANCESTRY TESTING 205

more people take such tests, these reference groups might change and as a result the ethnicity estimates for the same person might change across time. This does not devalue these tests as their results can indeed provide some valuable insights and information to people who may not know much about their ancestors.

2. *Not all your ancestors are represented in your DNA* (or, as I explain in Chapter 4, not all your genealogical ancestors are also your genetic ancestors). Therefore, the family history you grew up with may not be confirmed by your DNA testing results, not because it is not true but because your DNA does not include traces of that history. DNA must be considered as one source of information among several others—and perhaps not the most important one. Genealogical records, family stories and archives, may be more useful than DNA. As media and gender studies scholar Christina Scodari nicely put it: "genetic inheritance is negligible once one moves beyond third or fourth great-grandparents, meaning that cultural inheritance has a greater potential for lasting influence."[8] Indeed, it does. I am Greek because this is the ethnicity I was brought up into, because I speak Greek, because I was born in Greece, and because I follow the Greek customs and traditions. Whatever my DNA might say, it would likely not make much difference.

3. *The further back we go in time, the less we can find about our genetic ancestry and the more intertwined our family trees become.* Whether we claim descent from an important historical figure, from an important ancient civilization, or both, we should keep in mind that the number of our genealogical ancestors doubles after each generation as we go back in time. Eventually, we are all a huge family, and there is not much basis for claims about distinct ancestries—except for the very recent ones. As memory studies scholar Julia Creet suggested: "in the pursuit of descent and homelands, the genealogist uncovers quite the opposite: the dilution and discontinuities of bloodlines, the migrations of homelands, where these metaphysical desires are met by dissipation rather than concentration."[9]

4. *Whether we receive results about a unique or a mixed ancestry, we should keep in mind that these are the most probable estimates, and therefore relative.* For instance, if one receives results that one is 45% English, 30%

---

[8] Scodari, C. (2018). *Alternate roots: Ethnicity, race and identity in genealogy media.* Jackson: University Press of Mississippi, p. 115.

[9] Creet, J. (2020). *The genealogical sublime.* Amherst: University of Massachusetts Press, p. 23.

Italian, and 25% Greek, it should be clear that this is not a description of a person's genetic makeup, as if that was like a jigsaw puzzle consisting of pieces with different ancestry origins.[10] Rather, these percentages are a statistical estimate of that person's DNA similarity to the DNA of the people in the "English," "Italian," and "Greek" reference groups of the company. In addition, this does not mean that these portions of DNA are not similar to those of other reference groups; indeed, they could be. However, they happen to be more similar to one group rather than other groups, and it is only the former that is mentioned in the ethnicity estimate. Therefore, these percentages are relative rather than absolute.

5. *If you are told that you are "Greek," "Celt," "Viking," "Lombard," "Roman," or anything else, do not get overly enthusiastic.* You should remember that genealogical continuities across hundreds or thousands of years are hard, if not impossible, to establish, and that even if present and past populations are said to be related, this may not mean much—even if they have the same name. Other kinds of evidence such as archaeological and linguistic are important, too. Migration has always occurred throughout human history, even if this was done at a smaller scale than nowadays. Even if aDNA is used for such a conclusion, currently the inferences that can be made are limited. Continuity can be cultural, historical, or linguistic, but it does not have to be genetic.

If you have taken a test, you can reconsider their results after reading the present book. If you have not taken a test yet, keep in mind these issues before deciding why you really want to do it. But then are the tests any useful?

It Is DNA Family, Not Ancestry, Testing

What the tests can do really well is help you find relatives, that is, people with whom you have—relatively recent—common ancestors. This is based on the comparison of DNA segment by segment, and on the fact that people who have recent common ancestors will have DNA segments that are identical by descent (IBD, see Chapter 4). Let us then consider Ancestry's white paper on DNA matching, which is also recent, detailed, and clearly written, in order

---

[10] Corcos, A. F. (2018). *Three biological myths: Race, ancestry, ethnicity.* Tucson, AZ: Wheatmark.

FINDING MEANING IN OUR ANCESTRY TESTING   207

to see how this works.[11] As I explain in Chapter 4, IBD is about identifying DNA segments shared by pairs of individuals who have exactly the same sequences. When such shared DNA segments are found between two individuals, a "match" is said to have occurred. These segments are shorter the further back in time the common ancestor lived. The reason for this is that recombination of DNA resulting from crossing over (Box 4.2) in each generation results in smaller and smaller segments. This entails that a person will share relatively large segments with their parents, smaller ones with their grandparents, and even smaller ones with their great-grandparents. In other words, two individuals who have short IBD segments will probably be less closely related than two individuals who share longer segments.

A key problem in figuring out these relationships is to know from which parent each DNA segment came from. As we have seen, when DNA testing companies analyze the DNA of a person, they break it down to smaller segments, and they "scan" those segments for a several hundred thousand SNPs. Most SNPs are dimorphic, which means that they take one of two forms (e.g., A or C), which we can call alleles. If a marker has the same nucleotide twice (say A or C), this means that the individual has inherited the same allele from both of their parents. But if an individual has inherited an A and a C from each of their parents, we do not know which allele came from which parent. To figure this out, a process called "phasing" is used (Box 10.1). This is about finding out which nucleotides are "in phase," or have been inherited from the same parent. The best way to achieve this is to compare the DNA sequences of a person to those of their parents and figure out which alleles were inherited from which parent.

Once matches are found, another algorithm is used to estimate the possible relation between any two people in which those matches were found. In general, the principle followed is that more distantly related individuals (e.g., fourth cousins) are expected to inherit a smaller proportion of their DNA from common ancestors than individuals who are more closely related (e.g., first cousins). The expectation is therefore that more closely related individuals will have larger segments in common. Here is then how this works: First, each chromosome is divided into short segments, called "windows," which contain 96 SNPs (a number chosen to balance cost and accuracy). Then, for each pair of individuals "seed matches" are found, in which

[11] https://www.ancestry.com/dna/resource/whitePaper/AncestryDNA-Matching-White-Paper.pdf

Box 10.1 What Is Phasing?

Imagine that a person is found to have two DNA variants V1 and V2. If these are located each on one of the two homologous chromosomes, they must have been each inherited from each of that person's parents; that is, one variant will be of maternal origin, and the other will be of paternal origin. But it is also possible that V1 and V2 are located on the same chromosome. In that case, they will have been inherited from one of that person's parents, and they are said to be "in phase." Phasing is important because if we want to study the ancestry of a person, we have to know from which ancestor that person's variants have been inherited. The best way to achieve this is to have what is called a two parent–one child trio, or simply comparing the DNA of a person and their parents. The inclusion of siblings may increase the number of alleles that can be phased—in one study with data from families of two parents and four children, researchers were able to phase 98.8% of the alleles. If data from one or both parents are not available, other relatives might be used as proxy. Otherwise, it is possible to use algorithms to analyze simultaneously thousands of DNA sequences from unrelated individuals. Whereas in theory there should exist several combinations of the same alleles, what is found in practice is that particular combinations are found in many of the people studied. Eventually, the phase preferred is the one that results in two sequences on each of the chromosomes that are also observed in many other samples, under the expectation that short haplotypes are typically shared by many people in a large population. Once phasing is complete, the next step is to compare the DNA sequences of all individuals in the company's database two by two and look for identical sequences. This may sound simple, but it is extremely difficult as it entails comparing several hundred thousand markers for millions of customers, which could form trillions or pairs. Furthermore, the database changes continuously as more test-takers are added.

Roach, J. C., Glusman, G., Hubley, R., Montsaroff, S. Z., Holloway, A. K., Mauldin, D. E. (2011). Chromosomal haplotypes by genetic phasing of human families. *The American Journal of Human Genetics, 89*(3), 382–397.

FINDING MEANING IN OUR ANCESTRY TESTING 209

the alleles at all SNPs in one of each individual's two phased haplotypes are identical to those of the other individual. These "seed matches" are extended in both directions along the chromosome until either the end of the chromosome or a mismatch at both positions is reached. Then, the length of the candidate matching segment (in cM) is calculated, and the segments that are longer than 8 cM are retained (anything smaller is considered inaccurate). To this analysis another one is added for those cases where an individual has common DNA segments on both chromosomes as in the case of siblings (Ancestry calls this IBD2). This is done in order to be able to distinguish between full siblings and half siblings. In addition to estimating the length of each shared DNA segments, it is also possible to estimate the number of shared segments. Ancestry DNA has shown that eventually matches can be identified with a high level of accuracy. It is thus possible to distinguish between parents/offspring, identical twins, full siblings and half siblings.

So, to answer the main question in this section, what DNA ancestry tests do well is to find close relatives. The tests allow for the comparison of individuals to particular reference groups, which can reveal the relatedness between an individual and a particular reference population rather than another. In other words, the tests can show that a person is more likely to have a closer relation to some members of one group, say African Americans, than another, say European Americans. This is why many people have been able to find relatives, as well as to discover cases of nonbiological parents, especially fathers. This becomes possible not only on the platform of each company independently but also on platforms where it is possible to upload raw DNA data, the most well-known one being GEDmatch (https://www.gedmatch.com). However, caution is always required. Finding a match among three people does not entail that all three of them have the same common ancestor. Only if they all shared exactly the same IBD segment, could one infer that they are related. But even in that case, additional genealogical evidence would be required. DNA alone is not enough.

The available data make possible previously unthinkable achievements. In one study, starting from a pool of 86 million profiles from publicly available online data, it was possible to obtain population-scale family trees, including a single pedigree of 13 million individuals from almost every country in the Western world. The researchers noted that "We envision that this and similar large data sets can address quantitative aspects of human families, including genetics, anthropology, public health, and economics."[12] I think

---

[12] Kaplanis, J., Gordon, A., Shor, T., Weissbrod, O., Geiger, D., Wahl, M. et al. (2018). Quantitative analysis of population-scale family trees with millions of relatives. *Science, 360*(6385), 171–175.

that this is where the real value of the enterprise of ancestry testing lies: it helps one expand one's knowledge about family. There is no point in dealing with ethnicity or race. We had better begin from the most fundamental relations: parenthood and siblinghood. Then, rather than worry about identities that cannot be clearly delimited via DNA, such as ethnicity and race, we had better look for our closest relatives who can be established with sufficient certainty from the tests.

Is it time for the industry to rebrand itself? I think yes. I suggest that the whole enterprise should be called DNA family, not ancestry, testing.

Conclusion

In reaching the end of the present book, let me summarize the main argument that I have presented. Group identity is important, both for how one perceives oneself and for how one is perceived by others. Membership in a group, and the related feeling of belonging, provides security and solidarity. People subscribe to various social groups, but those related to ancestry are ethnic, national, and racial groups. These groups are socially constructed; as a result, they have no strict boundaries and are not internally homogeneous: ethnic groups can be defined on the basis of several different criteria, such as religion or language (e.g., the Lemba people think of themselves as Jewish, but look more similar to people in Africa rather than people in Israel); a person can acquire a new nationality during their life (e.g., a migrant who has lived for many years in a foreign country can often apply and acquire the nationality of the host country); and a race can include people with different visible characteristics (e.g., people from India and China are often described as "Asian"; not all people who identify as African Americans have very dark skin color). These social groups are often said to have had continuity across time historically, linguistically, culturally, and geographically. However, even though the labels we use to refer to past and current social groups may be the same, this does not mean that the referent is the same (e.g., Christian Orthodox religion is a defining feature of Greek ethnic identity today, but it was not one in ancient Greece—it did not even exist then). The categories around which these groups are formed are important for our lives, but the groups themselves are not clearly distinct or homogeneous.

However, we intuitively tend to essentialize these groups, that is, think of them as homogenous and clearly delimited from others with strict boundaries, due to some internal essences. When we think of ethnic, national, and racial groups, DNA can be perceived to serve as the placeholder for these essences. Given that our DNA molecules are approximately 99.9% identical, the perceived essential differences are restricted to the remaining 0.1%, which is still 4–5 million base-pairs. Population geneticists study these DNA sequences in order to figure out how different the various human

Ancestry Reimagined. Kostas Kampourakis, Oxford University Press. © Oxford University Press 2023.
DOI: 10.1093/oso/9780197656341.003.0011

212 ANCESTRY REIMAGINED

groups are, and which ones are more or less similar to which others. But to compare groups, these have to be somehow predefined, and researchers often tend to privilege those groupings that align with previously perceived, extant categories, such as races, nations, or ethnic groups. Even when the groups are not predefined but are supposed to emerge from the data, it seems that those groupings that align with previously perceived, extant categories are again privileged. Thus, racial, national, and ethnic groups are eventually naturalized, because their existence is seen as confirmed by the DNA data.

People living in the same continent are more likely to have recent common ancestors among themselves than with people living in other continents, and so the comparison of DNA will likely show more similarities among the former than between the former and the latter. But this is a difference of degree, not of kind. Human DNA variation is characterized by gradients/clines, not distinct groupings. People tend to mate with other people who live nearby, and so the closer two populations are found, the more similarly genetically (in terms of which alleles they have and in what frequencies) they are expected to be. The opposite is the case, obviously, for populations that are far apart. But because most of DNA variation is found among individuals rather than among populations, there is no way we can clearly assign individuals to groups on the basis of DNA data alone. Any such assignments are probabilistic. We create groups and we infer that some people are more likely to be assigned to one group rather than another. We based such inferences on the 0.1% of human DNA variation, perhaps overlooking the remaining 99.9%. In doing so, we also overlook that all humans have an ultimate origin and ancestry in Africa where our species evolved very recently in the evolutionary timescale.

The DNA variants that are used in the various human DNA variation studies and in DNA ancestry tests to differentiate between people of different ethnic groups are just found more frequently in one group rather than in another. There are no DNA variants that are found in one or another group only. Therefore, these DNA variants are in the best case indicia (see Chapter 3), not criteria, of ethnicity. They can be used to probabilistically assign individuals to (abstractly defined by the researchers or the companies) ethnicity groups, but they are not the reasons for which people have the ethnicity they have. This entails that whatever the results of a DNA test seem to suggest about ancestry should be considered carefully and thoughtfully. The DNA ancestry tests are intended to serve three general purposes, but they do not achieve all of them equally well.

CONCLUSION 213

1. *Identify particular individuals to whom a person is genetically related.* This is something that the tests can do very well. The comparison of DNA sequences in large databases can point to people with whom a person shares particular DNA segments that have been inherited from a common ancestor. One can thus find close or distant relatives with a high degree of accuracy. This is a reason to take a DNA ancestry test.

2. *Identify geographic areas where a person's ancestors lived.* Unfortunately, this is something that the tests cannot really do, except for some very recent ancestors. By finding relatives, or at least other people with similar DNA sequences in a particular geographically defined group, one can infer that a person's ancestors could have lived there. The fact that the members of a geographical group live today in a particular region makes it possible that their ancestors also lived there. But there is no way that we can establish this with certainty from DNA.

3. *Estimate the ancestry proportions of a person, for instance that a person has 50% European and 50% African American, or 40% Italian and 60% Irish ancestry.* Unfortunately, this is also something that the tests cannot do in any absolute term, but only in probabilistic terms. The tests cannot indicate that a person's DNA is 40% Italian and 60% Irish, as if it is a jigsaw puzzle. What the tests indicate is that 40% of a person's DNA is more similar to the reference group of the company described as Italian than any other of the groups; as well as that 60% of a person's DNA is more similar to the reference group of the company described as Irish than any other of the groups. But even this is something that is not well defined; it would only make sense if we were able to define what being 100% Irish or 100% Italian means, but this is not something we can specify accurately. However, there is a real value in such results, which lies in the fact that people who receive results about mixed ancestries realize that there are no pure ancestries, whatever these ancestries are. This will be the case for more and more people, as the reference groups and their geographic resolution increases.

Ancestry DNA test results can indeed provide some valuable insights and information to people who may not know much about their ancestors. But what the study of human DNA mostly shows is that we all are members of a huge, extended family, and not of genetically distinct ethnic groups. So what the tests are really good at is to point to genetic relatives, not ancestry. It is DNA family, not ancestry, testing. As geneticists Iain Mathieson and Aylwyn

Scally aptly put it: "most statements about ancestry are really statements about genetic similarity, which has a complex relationship with ancestry, and can only be related to it by making assumptions about human demography whose validity is uncertain and difficult to test."[1]

Both the findings of human population genetics and of DNA ancestry testing provide no evidence for racial or ethnic purity; in contrast, they show that human DNA variation is the outcome of migration and interbreeding. Genetic categories of whatever kind are not real; they are imagined and they cannot in any way substitute for historically constructed categories. DNA is not our essence and it cannot in any way indicate who we really are. Your ethnicity is not "written" in your DNA; it depends on your family, on where you were born and grew up, on your language and traditions. It depends on your social life, not on your DNA sequence.

---

[1] Mathieson, I., and Scally, A. (2020). What is ancestry? *PLoS Genetics*, *16*(3), e1008624.

Index

For the benefit of digital users, indexed terms that span two pages (e.g., 52–53) may, on occasion, appear on only one of those pages.

Tables, figures, and boxes are indicated by *t*, *f*, and *b* following the page number

Abu El-Haj, Nadia, 36–37
Accu-metrics, 21–22
admixture
  assumption of "pure" categories and
    populations, 146–47, 151–52
  defined, 146–47, 169–70
  legitimate use of concept, 153
  populations considered to have low
    levels of, 142–43
  problematic nature of, 146–47, 151–52
  reification hypothesis, 16–17
ADMIXTURE analysis, 136, 181–82,
  183–84
aDNA. *See* ancient DNA
adoption, 73, 91–92
Aegean Bronze Age, 176–84. *See also*
  Greek ancestry
  ancient DNA studies, 180–84, 180*t*, 183*t*
  periods and phases, 178–80
  regions, 178–79
affinal relatives, 74*f*, 74–75
  defined, 72–73
  genealogical ancestry, 77
African American Lives (television
  program), 54
African Americans. *See also* Black people
  effects of test results on self-perceptions
    and identity, 7–9, 17
  legitimate use of admixture
    concept, 153
  limits of YDNA and mtDNA
    analyses, 120–21
  self-reported ancestry vs. test results, 9–11
  slave trade and ancestry, 17, 54
  social construction of race, 62
  US Census descriptions, 62–63

Agro, Carly, 11–15, 13*t*
Agro, Charlsie, 11–15, 13*t*
AIMs (ancestry informative
  markers), 144–49
Alexandra (empress consort), 97–98
Alexei (tsesarevich), 97–98, 100–1
American Association of Biological
  Anthropologists, 57–58
American Society of Human
  Genetics, 57–58
Anastasia (grand duchess), 97–98, 100–1
ancestry
  caution needed in interpreting test
    results, 15, 125, 171, 204–6
  chicken-egg assignment of, 195, 197
  collective/global vs. individual/local, 23,
    109–10, 121–22, 195, 196, 197
  constructed rather than discovered, 35
  culture vs. genetics, 23, 194, 195,
    197–98, 205, 206, 214
  defined, 1–2, 101
  doubling of ancestors by generation, 76,
    83–84, 199–200, 205
  effects of study on self-perceptions and
    identity, 6–10, 17–22, 23, 175–76
  evolutionary nostalgia, 35
  family tree overlap and collapse, 76–77,
    103–4, 205
  focus on relatively tiny genetic
    differences, 109–10, 196–97, 212
  groupings as human inventions rather
    than natural, 23, 35, 63, 149–56, 172,
    187, 195, 197
  identity and, 24–25
  imagined fixed starting point, 102,
    104, 117

216 INDEX

ancestry (*cont.*)
  indigeneity, 25–27
  limited available information, 54,
    75–76
  monozygotic twins, 11–15, 14n.17, 23
  naturalizing categorization, 16–20,
    151–52, 211–12
  paradoxes underlying, 23, 195–98
  percentage and equivalency, 9, 78–81
  population genetics and individual-
    based genetics, 197
  relatedness vs. distinctiveness, 23, 123,
    143, 195, 196–97
  self-reported ancestry vs. test
    results, 22
  self-reported vs. test results, 9–11
  types of, 1–2, 78–81
  ultimate vs. proximate, 196
Ancestry.com and AncestryDNA
  addressing genetic ignorance in
    marketing, xi–xii
  DNA Journey, The (television
    program), x
  emphasis on kinship in marketing, 72
  essentialist marketing, 34
  exclusion criteria, 199–200
  genetic communities, 35–36
  limitations of ethnicity estimates,
    199–204
  number of profiles in database, xi
  surprising test results, 6–7
  television programs, 91–92
  testing methods and premises, 198–
    204, 206–9
  varied results among companies, 11–12,
    13t, 14–15
  white papers, 198–201, 206–7
ancestry informative markers
  (AIMs), 144–49
ancient DNA (aDNA), 149, 173–94, 206
  archaeologists and, 192–93
  "bottom- up" studies, 193
  challenges to study of, 173–75
  contamination with modern DNA, 174
  continuity between ancient and modern
    populations, 185–92
  damage and degradation over time, 173
  defined, 173

effects of study on self-perceptions and
    identity, 175–76
  fundamental questions about, 192–93
  Greek ancestry case study, 176–92
Anderson, Benedict, 39–40
Angelou, Maya, 54
answer seekers, 6, 91, 92–93, 101
Antetokounmpo, Giannis, 68–69
Appiah, Kwame Anthony, 24–25, 50–51
ascertainment and ascertainment
    bias, 147–48
Asian Americans, 17
assisted reproduction, 6–7, 73
Aurelie (*The DNA Journey*
    participant), 102
autosomal DNA
  comparing variation of populations, 112
  "deep" ancestry, 120–21
  qualitative and qualitative differences
    between populations, 112
  representation of ancestry in,
    83–84, 90
  YDNA and mtDNA vs., 86, 87b, 90, 94–
    95, 99b, 104n.4, 117, 120–21, 196
autosomes
  crossing over, 80b–81b
  defined, 79b
  pseudo-autosomal regions, 87b
Avdonin, Alexander, 97–98
avid genealogists, 6
Axmann, Artur, 94

Barbujani, G., 124t
base calling, 8b
base pairs
  centiMorgans, 84–85
  focus on relatively tiny genetic
    differences, 211–12
  length of DNA, 5b
  mutation rate, 99b, 108n.18
belonging
  constructed rather than discovered, 35
  DNA ancestry testing and, 35–36
  effects of study of aDNA on self-
    perceptions and identity, 175–76
  emotional attachment, 25–26
  emphasis on kinship in test
    marketing, 72

INDEX   217

exclusion and, 25
genealogical imagination, x–xi
group identity, 24–25, 211
identity and, 27–28
indigeneity, 26–27
kinds of, 153–54, 171, 175–76
language and, 40, 47–48
sense of familiarity, 25
Ben (sperm donor), 6–7
Benson, Philip, 34–35
Bettinger, Blaine, 15
Beyonce, 60–62
biogeographical ancestry (BGA), 144–49
biological race
  academic views on nature of race, 55–57
  biogeographical ancestry, 144
  clustering and, 134–35
  defined, 4
  distinct populations vs., 149, 151
  essentialism, 51–52
  persistence of flawed idea, 137–43
  rejection of biological conceptions of
    race, 56–58
  science communication on DNA
    and, 22
  scientific racism, 53–54
BioMe, 160–61, 161f
Birney, Ewan, 155–56
Black people. See also African Americans
  effects of test results on self-perceptions
    and identity, 18
  essentialism, 51–52
  intersections of gender and race, 28–29
  "one drop rule," 62
  social construction of category, 60–62
  US Census descriptions, 62–63
Blaschke, Hugo Johannes, 94
Bliss, Catherine, 153
Boas, Franz, 54–55
Bolnick, Deborah, 132–33
Bormann, Martin Ludwig, 93–95
  background of, 93
  disappearance of, 93
  DNA analysis, 94–95
  skeletal remains, 94–95
Braun, Eva, 94
brow ridge (supra-orbital ridge) and skull
  comparisons, 52–53, 53f

Brubaker, Rogers, 16, 67–68, 148–49
Burton, Elise, 36–37

Canadian Broadcasting
  Corporation, 11–12
Cann, Rebecca, 104
Carlos (The DNA Journey participant), 102
Carr, John, 96
Carr, Martha Jefferson, 96
Carr, Peter, 96
Carr, Samuel, 96
Case Western Reserve University, 14
Caucasoids/Caucasians, 155. See also
  White people
Cavalli-Sforza, Luigi Luca, 128–29,
  168–70, 177–78, 199–200
Center for African and African American
  Research, 7–9
centiMorgans (cM), 84–85, 84n.23
Centre d'Étude du Polymorphisme
  Humain (CEPH; Center for the Study
  of Human Polymorphisms), 38
Charlemagne, 76–77, 86–90
Cheddar man, 175–76
cherry-picking of data, 164–65
chromosomes
  crossing over, 80b–81b, 83
  DNA ancestry testing methods, 207–9
  inheritance of, 78–86, 82f
  overview of, 79b
  pseudo-autosomal regions, 87b
  YDNA, 87b
Church of Jesus Christ of Latter-day Saints
  (Mormons), 75–76
civic nations, 30, 32
Clemente, F., 186–87, 189–90
clines and clinal variation, 128–29,
  133–35, 151–52, 156–62, 212
clustering, 137–38, 143
  caste or religious groups and, 162
  continuum of variation vs., 130–
    36, 160–61
  exclusion criteria, 164
  geography and, 162
  sampling bias, 158, 159f, 167–68
cM (centiMorgans), 84–85, 84n.23
CNRS (National Center of Scientific
  Research), 186–87

218 INDEX

CNVs (copy-number variants), 33*b*
coalescence time, 104, 106–7
Cobb, Craig, 7–9, 21
Collins, Jim, 34–35
confirmation bias, 186
consanguineal relatives, 72–73, 74*f*, 74–75
continental and subcontinental regions
   and groups, 161*f*, *See also* clustering
  ambiguity in classifying
   populations, 153–54
  biological race, 137–38
  clustering humans into groups, 133–
   34, 135–36
  common ancestors, 212
  DNA testing pinpointing ancestry to,
   11, 63–64, 69–70, 123–25
  HapMap Project, 109
  identification of races and ethnicities
   with, 11, 48, 58–59, 162
  medical research, 126
  mtDNA haplogroups, 121
  overlapping variation among,
   123, 124*f*
  perceptions of national identity, 41–43
  predefined populations, 127–28
continental belonging, 175–76
Coon, Carleton, 55
Coop, Graham, 86–90, 191–205
Copeland, Libby, 6, 34–35
copy-number variants (CNVs), 33*b*
cosmopolitanism, 29, 59–60
Creet, Julia, 205
Crenshaw, Kimberle, 28–29
crossing over, 80*b*–81*b*, 83
cultural ancestry, 1, 2
cultural diffusion, 128

Darwin, Charles, 105
Dass, Angélica, 59–60
demic expansion, 128
Denisovans, 49, 114, 115, 116–17, 173,
  175, 192
DeSalle, Robert, 147–48, 164–65
Descent of Man, The (Darwin), 105
diploid organisms and cells, 79*b*,
  180, 185–86
DNA. *See also* ancient DNA; DNA
  ancestry testing; DNA replication;

DNA sequencing; human
  evolutionary genetics; mutations
  mtDNA, 88*b*
  overview of, 5*b*
  YDNA, 87*b*
DNA ancestry testing
  affordability, 10–11
  ancient DNA, 193–94
  biological markers vs. non-biological
   markers, 196, 197–98
  caution needed in interpreting results,
   15, 125, 171, 204–6
  changes to ethnicity estimates over
   time, 202, 204–5
  chicken-egg assignment of ancestry,
   195, 197
  confusion between race, ethnicity, and
   geography, 11
  confusion in test result labels, 11, 41
  continuity across time, 185–92, 206
  criteria vs. indicia for ethnicity, 69–
   71, 212
  dealing with consequences of slave
   trade, 54
  "deep" ancestry, 121–22
  effects of results on self-perceptions and
   identity, 6–10, 17–22, 23
  emphasis on kinship in marketing, 72
  essentialism, 18–20, 20n.27, 34,
   69–70
  estimating ancestry proportions, 213
  ethnicity estimates, 3, 6–7, 11–12, 15,
   63–64, 146, 198–204
  evolutionary nostalgia, 35
  exclusion criteria, 199–200
  extent of genetic ancestry information
   provided by, 90
  genetic characteristics vs. ethnic
   identity, 66, 69–70
  genetic communities, 35–36
  genetic histories, 107–8
  identifying geographic areas where
   ancestors lived, 213
  identifying relatives, 213
  imagined communities and
   categories, 40–41
  imagined fixed starting point of
   ancestry, 102, 104

labels referring to geographic regions, 202

limitations of, 92–93, 97, 101, 198–204

marketing, xi–xii, 34, 72

modern populations, 145–46, 171, 204–5

monozygotic twins, 11–15, 14n.17, 23

motivations for testing, xi–xii, 91–92

multidimensional concept of, 195, 196

naturalizing categorization, 16–20, 211–12

not really about ancestry, 204–5, 206–10, 213–14

number of profiles in databases, xi

oversimplification of complex past, 103–4

paradoxes underlying ancestry, 23, 195–98

parenthood and filiality, 73–74, 209–10

patterns of variation in populations, 123–25

percentage and equivalency, 9

perceptions of and attitudes toward, 16–17

population genetics and individual-based genetics, 125

privileging of distinctiveness, 195, 196–97

privileging of genetic ties, 103–4

purposes served by, 212–13

reference groups, 10–11, 12–15

relatedness of contemporary people through recent common ancestry, 145–46, 171

relativity vs. absoluteness, 201–2, 203, 205–6

reliance on evidence from additional sources for interpretation, 92–93, 95, 97, 101

representation of ancestry in DNA, 78–84, 82f, 90, 205

representativeness of reference groups, 10–11, 133–34

sample size, 111

selective interpretation for political purposes, 21–22, 185

self-reported ancestry vs. test results, 9–11, 22

sense of belonging, 35–36

studies about how people interpret results, 16–22

study of human DNA variation at population level, 108–14

suggested rebranding to DNA family testing, 206–10

terminology for results, xi–xii

types of test takers, 6

varied results among companies, 11–15, 13t, 14n.17, 23

DNA Journey, The (television program), ix–x, 102–3

DNA medical and health testing, 4–6, 126–27

DNA replication, 33b, 79b, 80b

DNA sequencing
  Ancestry test sample processing procedure, 198–99
  coverage or depth, 8b
  data vs. evidence, 14
  errors, 8b
  inferences based on sequence data about individuals, 14
  next generation sequencing, 8b, 173, 173n.1
  overview of, 8b
  variation in sequence, 33b

Dobzhansky, Theodosius, 54–55, 56

Donovan, Brian, 19–20

Elizabeth II (queen), 98

Ellaha (*The DNA Journey* participant), 102–3

epicanthic fold, 58–59

essentialism, 31–36, 69–70
  ancient DNA studies, 187–88
  "blood" relations, 75
  defined, 31
  DNA essentialism, 32, 32n.14
  enhancing cohesion and exclusion, 32–34
  ethnic groups, 66–67, 66–67n.38
  ethnic nation concept, 30
  genetic characteristics vs. ethnic identity, 66
  genetic essentialism, 32n.14
  genetic ethnicities, 40–41

220 INDEX

essentialism (*cont.*)
  group identity, 211–12
  groupism, 67–68
  imagined communities and
    categories, 40–41
  interrelated beliefs comprising, 31,
    32, 34, 51
  level of genetics and genomics
    knowledge and, 19–20, 20n.27, 22
  perceptions of national identity, 41–
    48, 44f
  primordialism, 66–67
  "pure" categories and
    populations, 151–52
  racial essences, 51
  scholars' views, 56–57, 57n.18
  stereotyping, 32
  study of test results and, 19–20
ethnic identity, x–xi
  ancient Greece, 191–92
  assertion of, 68–69, 191–92
  as contingent phenomenon, 191–92
  culture vs. genetics, 70–71, 191–92, 194
  disconnect between DNA and, 191–
    92, 194
  genetic characteristics vs., 66, 69–70
  immigration and globalization, 65
  imposition of vs. choice of, 65–66
  primordialism vs.
    instrumentalism, 66–67
ethnicity, 64–69
  acquiring independently of genetics,
    34–35, 194, 214
  African ethnic categories, 116
  caution needed in interpreting test
    results, 15
  confusion in test result labels, 11
  criteria vs. indicia for, 69–71, 212
  defined, 2
  definitional vs. operational sets of
    attributes, 69
  effects of test results on self-perceptions
    and identity, 6–10
  ethnicity estimates, 3, 6–7, 11–12, 15,
    63–64, 146
  ethnic nation concept, 29–30
  ethnic prejudice, 65

ethnocentrism, 65
  etymology of word, 64–65
  focus on, 3, 64
  genetic ethnicities, 40–41
  geneticists' transformation of identities
    into, 37–38
  groupism, 67–68
  identity and, 24–25
  imposition of ethnic identity, 65–66
  monozygotic twins, 11–15, 14n.17, 23
  naturalizing categorization, 16–
    20, 211–12
  as permeable, 68–69
  race and nationality vs., 3, 64
  self-reported vs. test results, 10–11,
    160–61, 161f
  social construction of ethnic
    groups, 211
  varied test results among companies,
    11–15, 13t
ethnic nations, 29–30, 32, 64–65
ethnic/racialized belonging, 175–76
eugenic policies, 54–55
European Americans. *See also*
    White people
  self-reported ancestry vs. test
    results, 9–11
  terminology, 155
Evans, Arthur, 178–79
evolutionary nostalgia, 35
exclusion criteria, 133–34, 147, 162–63,
    164–65, 164t, 172, 199–200

Falk, Raphael, 36–37
FamilyTreeDNA
  number of profiles in database, xi
  promoting DNA testing, xi–xii
  study of results and essentialist view of
    race, 19–20
  varied results among companies, 11–
    12, 13t
family tree overlap and collapse, 76–77,
    103–4, 205
Feodorovna, Maria, 98–100
fixity and stability
  essentialism, 187–88
  ethnicity, 66–67, 68–69

INDEX 221

imagined fixed starting point of
ancestry, 102, 104, 117
fossils, 49, 114, 115, 173
frappe software, 135
Freeman, Morgan, 54
Friedrich Schiller University, 57–58
Fuentes, Daisy, 91–92
Fullwiley, Duana, 147

Gannett, Lisa, 144
Gates, Henry Louis, Jr., 7–9, 54, 120–
21, 145–46
Geary, Patrick, 171–72, 193
GEDmatch, 209
gender, 28–29, 32n.14, 87*b*
genealogical ancestors and ancestry, 72–90
consanguineal vs. affinal relatives, 72–
73, 74–75, 74*f*
defined, 2, 78–81
doubling of ancestors by generation, 76,
83–84, 199–200, 205
family tree collapse, 76–77
genealogical ancestors not always also
genetic ancestors, 82, 82*f*, 83–84, 90,
91, 107, 201–2, 205
genealogical distance, 73–74, 77–78
generation time, 73–74
genetic ancestry vs., 2, 78–84, 82*f*,
84n.20, 90, 91, 103–4, 107, 201–2, 205
"ghost" ancestors, 83–84
kinship, 72–73
limited available information, 75–76
parenthood and filiality, 73–74
representation of ancestry in DNA, 78–
84, 82*f*, 90
genealogical imagination, x–xi, 102
generation time (generation interval),
73–74, 76, 103
genetic ancestors and ancestry, 72–90
"blood" relations, 73, 75, 76–77
chromosomal inheritance, 78–86, 82*f*
common genetic ancestors, 86–90,
102, 103
consanguineal vs. affinal relatives, 72–
73, 74–75, 74*f*
cultural ancestry vs., 2
defined, 1, 78–81

family tree collapse, 76–77
genealogical ancestors not always also
genetic ancestors, 82, 82*f*, 83–84, 90,
91, 107, 201–2, 205
genealogical ancestry vs., 2, 78–84, 82*f*,
84n.20, 90, 91, 103–4, 107, 201–2, 205
geographical ancestry vs., 2
"ghost" ancestors, 83–84
how far back genetic ancestry can be
identified, 86–90
kinship and, 72–73
medical research, 126–27
parenthood and filiality, 73–74
privileging of genetic ties, 72–73, 76–77,
90, 103–4
race vs., 126–27
representation of ancestry in DNA,
78–84, 90
genetic communities, 35–41
French DNA as "national patrimony," 38
genetic ethnicities, 40–41
geneticists' transformation of identities
into ethnicities, 37–38
Jewish people, 36–37
nation as imagined community, 39–40
genetic ethnicities, 40–41, 64, 69–70
genetic histories, 107–8, 172, 176–78
Genetic History of Greece, The
(Triantaphyllidis), 177–78
genetic ignorance, xi–xii
genetics/genomics knowledge and
education, 19–20, 22
Genographic Project (GP), 109
genome, 5*b*, 8*b*
geographical ancestry
confusion in test result labels, 11
defined, 1
genetic ancestry vs., 2
Georgij (grand duke), 98–100
Gerstein, Mark, 12–14
"ghost" ancestors, 83–84
Gibbons, Ann, 185
globalization, 65, 170
Goddard, Trisha, 7–9
Goodman, Alan, 1–2
GP (Genographic Project), 109
Graves, Joseph, 1–2

222 INDEX

Greek ancestry
  ancient DNA studies, 180–92, 180*t*, 183*t*
  ancient periods and phases, 178–80
  assertion of ethnic identity, 68–69
  continuity between ancient and modern
    Greeks, 177–78, 182, 184–92
  criteria vs. indicia for ethnicity, 69
  culture vs. genetics, 194
  Greece/Asia Minor population
    exchange, 163, 163n.32
  imposition of ethnic identity, 65–66
  map of ancient Greece, 177*f*
  stereotypes, 27–28, 32
groupings and groups. *See also* continental
    and subcontinental regions
    and groups
  assumption of "pure" categories and
    populations, 146–47, 151–52, 169–
    70, 213
  essentialism, 211–12
  group identity, 211
  as human inventions rather than
    natural, 23, 63, 149–56, 172, 187,
    195, 197
  imagined communities and categories,
    39–41, 214
  importance of, 211
  naturalizing categorization, 16, 151–
    52, 211–12
  not distinct or homogenous, 211–12
  predefined, 211–12
  privileging of, 211–12
  social construction of, 211
groupism, 67–68
Guibernau, Montserrat, 25

Hall, Jonathan, 190–92
Hamilakis, Yannis, 188
Hammer, Michael, 106–7
haplogroups, 118–20, 119*f*, 121
haploid organisms and cells, 79*b*, 87*b*, 118
Haplotype Map (HapMap) project, 109,
    110*t*, 139–41, 143, 159–60
haplotypes, 118, 119*f*, 140*b*, 198–99, 207–
    9, 208*b*
Harvard University, 7–9
health and medical research, 4–6, 126–27
Heidi (*A New Leaf* participant), 91–92

Hemings, Betty, 95–96
Hemings, Beverley, 95–96
Hemings, Eston, 95–97, 101
Hemings, Harriet, 95–96
Hemings, Madison, 95–96
Hemings, Sally, 95–97
  DNA analysis of descendants, 96–97
  relationship with Thomas
    Jefferson, 95–96
Hess, Rudolf, 93–94
heteroplasmy, 98–100, 99*b*
HGDP (Human Genome Diversity
    Project), 109, 110*t*, 121, 134, 135,
    167–70, 168*f*
HGP (Human Genome Project), 109n.20
Hispanic and Latino people
  effects of test results on self-perceptions
    and identity, 17
  perceptions of and attitudes toward
    testing, 17
  self-reported ancestry vs. test results, 10
Hitler, Adolf, 93–94
hobbyists, 6
Hochschild, Jennifer, 17
Holocaust, 54–55
homogeneity
  essentialism, 18–19, 31, 51–52, 211–12
  ethnicity, 64–65
  Greek aDNA studies, 182, 183–
    84, 187–89
  groupism, 67–68
  nationality, 25–26, 31
  race, 51–52
  social construction, 211
Homo sapiens, 49–50
Humanæ (Dass), 59–60
Human Diversity (Murray), 137
human evolutionary genetics, study
    of, 123–27
  absolute indication of geographical
    region or population, 147–48
  assumption of "pure" categories and
    populations, 146–47, 151–52, 169–
    70, 213
  "bottom-up" community-based
    studies, 193
  choice of terminology and
    language, 155–56

clines and clinal variation, 128–29, 133–35, 151–52, 156–62, 212
clustered vs. continuous DNA variation, 130–37
criteria vs. indicia for ethnicity, 69–71, 212
demic expansion vs. cultural diffusion, 128
differences of degree, not kind, 212
"dilution" of inherited and shared DNA across generations, 82–84, 171
DNA variation overview, 33*b*
exclusion criteria, 133–34, 147, 162–63, 164–65, 164*t*, 172
fluctuation in, 110–11
history and geography of populations reflected in DNA variation, 127–30
human genome projects, 109–10
importance of variants used for comparison, 114
inheritance of segments vs. of sequences, 86n.25
isolation by distance, 156–57
measurement and expression of DNA variation, 112n.30
method-driven findings, 135–36
migratory processes and related founder effects, 157
natural selection and adaptation, 162
"out of Africa" theory, 115–16
percentage of DNA shared with all other humans, 109–10, 143, 149, 196–97, 211–12
perceptions of DNA, 4
population genetics and individual-based genetics, 197
population-level study of DNA variation, 108–14
population structure, 130–33
principal component analysis, 128–30
qualitative and qualitative differences, 112–13, 113*t*, 113*f*
representation of ancestry in DNA, 78–84, 90
representativeness in sampling, 133–34, 147, 165–67, 168–70, 172, 185–86
sample size, 168–69, 185–86

sampling procedures and bias, 156, 158, 162–72, 164*t*, 166*t*, 185–86
selective sweeps, 139–41, 140*b*, 142
variation within minor portion of genome, 109–10
variation within regional groups rather than among them, 123, 124*f*, 124*t*
weighting of subsamples, 165, 167
Human Genome Diversity Project (HGDP), 109, 110*t*, 121, 134, 135, 167–70, 168*f*
Human Genome Project (HGP), 109n.20
hybridization, 149–52, 153–54, 179–80
hypervariable regions (HVRs), 88*b*, 94–95, 98–100
"hypo-descent rule" ("one drop rule"), 62, 62n.28

Iceman, 105–6
identical by descent (IBD) segments, 84–90, 85*f*, 199–200, 206–9
identity. *See also* ethnic identity
acquiring independently of genetics, 34–35
ancestry and, 24–25
attributes and identifying features, 24
belonging and, 27–28
effects of ancestry research on, 6–10, 17–22, 23, 175–76
ethnicity and, 24–25
group identity, 24–25, 211
intersectionality, 28–29
nationality and, 24–26, 27–30, 39
perception and relationships with others, 24–25
race and, 24–25
sense of belonging, 24–26, 27–30
Ignatieff, Michael, 29, 47–48
immigration and immigrants. *See also* migration
essentialist perceptions of national identity, 41–45, 44*f*
ethnic identity, 65
immigrants and children of immigrants becoming true nationals, 41–48
indigeneity, 26–27
Inclusiveness of Nationalities, The (survey), 41–46, 44*f*

224 INDEX

indigeneity, 25–27
Indigenous peoples. *See also* Native Americans
 caution needed in terminology, 171
 disappearance due to genetic admixture, 27
 Euro-centric justifications for research into DNA of, 27
 indigeneity, 25–27
 lack of consultation and community engagement in studies, 27
 White French Canadians and claims of Indigenous identity, 21–22
individuation, 127–28, 149–56
Inouye, Michael, 155–56
instrumentalism, 66–67
intersectionality, 28–29
IPSOS, 41–42
isolation by distance, 156–57

James, Lebron, 60–62
Jay (*The DNA Journey* participant), ix–, 18–19, 66, 69–70, 102, 201
Jefferson, Field, 96, 97
Jefferson, Martha Wayles, 95–96
Jefferson, Randolph, 97
Jefferson, Thomas, 95–97, 101
 background of, 95–96
 DNA analysis of descendants, 96–97
 relationship with Sally Hemings, 95–96
Jena declaration on race, 57–58
Jewish ancestry, 6–7, 28–29, 34–35, 36–37, 65–66
Jobling, Mark, 125
Jones, Quincy, 54
Jordan, Daniel P., 97
Jorde, L. B., 124*t*

Karen (*The DNA Journey* participant), ix, 18–19, 66, 69–70, 102
Kempka, Erich, 94
Kennewick Man, 175–76
kinship
 assisted reproduction and adoption, 73
 defined, 72–73
 importance attributed to, 72–73
Krause, Johannes, 173

Krimbas, Costas, 177–78
Kroeber, Alfred, 54–55

language
 belonging and, 40, 47–48
 development of national consciousnesses, 39–40
 essentialist view of nationality, 46–48
 ethnic identity, 2, 65–66
 ethnic vs. civic nations, 29–30
 Greek aDNA studies, 176–78
 language, 2
 sense of belonging and, 47–48
"Language of Race, Ethnicity, and Ancestry in Human Genetic Research, The" (Birney et al.), 155–56
Latino people. *See* Hispanic and Latino people
Lawson, Daniel John, 136
Lazaridis, Iosif, 178–79, 182, 185–86, 187, 188, 190–91
Lee, Christopher, 76–77, 86–90
Leroux, Darryl, 21–22
Lewontin, Richard, 123, 124*t*, 165
Li, J. Z., 135, 137–38
Linnaeus, Carl, 49–50
Living DNA, 11–12, 13*t*
Lost Family, The (Copeland), 34–35
Louise of Hesse-Cassel (queen), 98
Lowe, Rob, 72

Malaspinas, Anna-Sapfo, 182
male-specific region of the Y chromosome (MSY), 87*b*, *See also* YDNA
Mallon, Ron, 60–62
Maran, Joseph, 188, 191
Maria (grand duchess), 97–98, 100–1
Marketplace (television program), 11–12
Marks, Jonathan, 53–54, 146–47
Mathieson, Iain, 213–14
M'Charek, Amade, 152–53
meiosis, 78–81, 80*b*–81, 83, 87*b*
melanin, 58*b*, 61*b*
Menozzi, Paolo, 128
methodological nationalism, 37–38
Middleton, Guy D., 188–89

INDEX 225

migration. *See also* immigration and immigrants
  assumption of limited migration, 170, 199–200, 206
  civic nations, 30
  DNA variation and, 110–31, 157, 201–2, 214
  Greece/Asia Minor population exchange, 163
  "out of Africa" theory, 116–17
Millennium Pharmaceuticals Inc., 38
mitochondria, 88*b*
mitochondrial DNA (mtDNA)
  autosomal DNA vs., 94–95, 99*b*, 104n.4, 117, 120–21
  coalescence time, 106–7
  "deep" ancestry, 117–21
  haplogroups, 118–20, 119*f*, 121
  haplotypes, 118, 119*f*
  Iceman case, 105–6
  inheritance of, 86, 88*b*, 89*f*
  limitations of focus on, 120–21
  Martin Bormann case, 94–95
  "mitochondrial Eve," 104–7, 104n.4
  "out of Africa" theory, 105
  overview of, 88*b*
  representation of ancestry in, 90
  Romanov case, 98–100
  YDNA vs., 88*b*, 196
"Mitochondrial DNA and Human Evolution" (Wilson, Cann, and Stoneking), 104
"mitochondrial Eve" (mt-Eve), 104–7, 104n.4
mitosis, 80*b*
Modern Evolutionary Synthesis, 54–55
Momondo, x
monozygotic twins, 11–15, 14n.17, 23
Mormons (Church of Jesus Christ of Latter-day Saints), 75–76
Morning, Ann, 3, 56–57
MSY (male-specific region of the Y chromosome), 87*b*, *See also* YDNA
mtDNA. *See* mitochondrial DNA
mt-Eve ("mitochondrial Eve"), 104–7, 104n.4
"multiregional" theory of human evolution, 105, 106–7

Murray, Charles, 137–38, 139–41, 142–43
mutations
  clustering, 162
  haplotypes, 118
  heteroplasmy, 99*b*
  mtDNA, 99*b*, 104–5
  "out of Africa" theory, 115–16
  overview of, 33*b*
  rate of, 33*b*, 87*b*, 99*b*, 108–9, 108n.18, 171
  selective sweeps, 140*b*
  skin color, 61*b*
  YDNA, 87*b*
MyHeritage
  number of profiles in database, xi
  promoting DNA testing, xi–xii
  varied results among companies, 11–12, 13*t*, 14–15
My True Ancestry, 193–94

Nash, Catherine, 118–20
national belonging, 25–26, 27–30, 39, 175–76
National Center of Scientific Research (CNRS), 186–87
nationality
  acquiring independently of genetics, 34–35, 46
  conceptions of, 29–30
  defined, 2
  essentialist perceptions of national identity, 41–48, 44*f*
  ethnicity vs., 3
  genetic nationalism, 37–38
  identity and belonging, 24–26, 27–30, 39
  immigrants and children of immigrants becoming true nationals, 41–48
  important features for being a true national, 46–47
  interrelated beliefs comprising essentialism, 31, 32, 34
  intersectionality, 28–29
  maintaining one's own nationality abroad, 41–48
  methodological nationalism, 37–38
  multiethnic nations, 47–48
  nationalism, 29

226 INDEX

nationality (*cont.*)
  nation as imagined community, 39–40
  naturalizing categorization, 211–12
  notion of indigeneity, 25–27
  prioritizing, 29
  social construction of, 211
  stereotyping, 27–28, 32
Native Americans. *See also* Indigenous
    peoples
  effects of test results on self-perceptions
    and identity, 17–18
  Euro-centric justifications for research
    into DNA of, 27
  modern human migration into
    Americas, 116–17
  White French Canadians and claims of
    Indigenous identity, 21–22
naturalizing categorization, 16, 151–
    52, 211–12
natural selection and adaptation, 110–11,
    130–31, 137, 138–42, 140*b*, 149–50,
    157–58, 162
Nature Genetics, 153
NBC, 91–92
Neanderthals, 49, 114, 115, 116–17, 149–
    51, 153, 173, 174–75
Nelson, Alondra, 54
New Leaf, A (television program), 91–92
next generation sequencing (NGS), 8*b*,
    173, 173n.1
Nicholas I (emperor), 100
Nicholas II (tsar), 97–101
nonrecombining region of the Y
    chromosome (NRY), 87*b*, *See
    also* YDNA
Nott, Josiah Clark, 52–53, 53*f*
Novembre, John, 128–30, 156–57, 162–63,
    164–65, 164*t*, 166*t*, 167–68
nucleotides. *See also* single nucleotide
    polymorphisms
  DNA sequencing, 8*b*
  haplotypes, 118
  overview of, 5*b*
  phasing, 207
  Romanov case, 98
  whole-genome sequencing, 147–48

Obama, Barack, 62

Oikkonen, Venla, 35, 175–76
Olga (grand duchess), 97–98
1,000 Genomes Project (1KGP), 109–10,
    110*t*, 118–20, 121, 143
"one drop rule" ("hypo-descent rule"),
    62, 62n.28
Orlando, Ludovic, 186–87
"out of Africa" theory of human evolution,
    105, 106–7, 114–17, 212
  "deep" ancestry, 121–22
  dispersal from Africa, 116–17
  ethnic categories, 116
  haplogroups and, 118–20
  inferring African ancestry, 116
  main phases of, 114–15
Oxford Ancestors, 106n.8

Pääbo, Svante, 133–35, 158, 173
PAAs (population associated alleles), 144
paleogeneticists, 192–93
Panofsky, Aaron, 21, 153
Papageorgopoulou, Christina, 182
PARs (pseudo-autosomal regions), 87*b*
PCA (principal component analysis),
    128–30, 135, 160–61, 161*f*, 162–
    63, 181–82
personal belonging, 175–76
PEW Research Center, 46–48, 46n.29
phasing, 207, 208*b*
Phelan, Jo, 16–17
Philip (prince), 98
Phillips, Andelka, 6
Piazza, Alberto, 128
Popejoy, Alice, 154–55
population associated alleles (PAAs), 144
populations
  as abstractions, 152–53, 156
  ambiguity in classifying, 153–54
  assignment of individuals to, 127–
    28, 197
  assumption of "pure" populations, 146–
    47, 151–52, 169–70, 213
  considered to have low levels of
    admixture, 142–43
  continuity between ancient and
    modern, 185–92
  distinct populations vs. biological race,
    149, 151

genetic constitution of, 111
genetic definition of, 110–11
importance of variants used for comparison, 114
individuating, 149–56
patterns of variation in, 123–25
population genetics and individual-based genetics, 125, 197
population structure, 130–33, 176–78
predefined, 127–28, 130
qualitative and qualitative differences between, 112–13, 113f, 113t
sample size and representativeness, 111
variation in designation among studies, 154–55
population-specific alleles (PSAs), 144–45
primordialism, 66–67
principal component analysis (PCA), 128–30, 135, 160–61, 161f, 162–63, 181–82
Pritchard, Jonathan, 130–32, 134–35, 138–42, 143
Przeworski, Molly, 152
PSAs (population-specific alleles), 144–45
pseudo-autosomal regions (PARs), 87b

race
academic views on nature of, 55–57, 57n.18
biological characteristics distinguishing non-biological races, 58–60
conceptions of, 3–4
confusion in test result labels, 11
defined, 2
effects of test results on self-perceptions and identity, 16–22
essentialism, 18–20, 51, 57n.18
ethnicity vs., 3
eugenic policies, 54–55
genetic ancestry vs., 126–27
identity and, 24–25
lack of scientific support for biological race, 49–57
medical research, 126–27
naturalizing categorization, 16, 211–12
philosophical arguments about, 60n.26
racial fixation, 50–51
racialism, defined, 53–54

racism, defined, 53–54
reification hypothesis, 16–17
rejection of biological conceptions of, 56–58
science communication on DNA and biological race, 22
scientific racism and early scientific concepts of, 49–50, 52–56
social construction of, 56–64, 211
use and misuse of concept in medicine, 126–27
"Race in America" series, 7–9
Raff, Jennifer, 155–56
Ralph, Peter, 86–90, 191
regional belonging, 175–76
Reich, David, 149–51
Roberts, Dorothy, 146–47
Romanovs, 97–101
background of, 97–98
discovery of graves, 97–98
DNA analysis, 98–100
Rosenberg, Noah A., 124t, 132–33, 134–35, 137–38, 167–68
Roth, Wendy, 17–18, 19–20
Rutherford, Adam, 155–56
Rwandan Hutu/Tutsi conflict, 65
Ryabov, Gely, 97–98

Sabeti, Pardis, 141–42
Sand, Shlomo, 65–66
Scally, Aylwyn, 155–56, 213–14
Schliemann, Heinrich, 178–79, 188–89, 190–91
Science, 185
Scodari, Christina, 205
selective sweeps, 139–41, 140b, 142
Sen, Maya, 17
Serre, David, 133–35, 158
sex chromosomes, 79b
SGDP (Simons Genome Diversity Project), 109, 110t
Shapiro, Dani, 6–7, 73, 201–2
Shapiro, Michael, 6–7
Shapiro, Susie, 6–7
Shriver, Mark, 144–45
sickle-cell anemia, 126, 126n.5
Simons Genome Diversity Project (SGDP), 109, 110t

228 INDEX

Simpson, Bob, 40
single nucleotide polymorphisms (SNPs),
    84–85, 111, 121, 129–30, 135, 144–
    45, 147–48
  Ancestry test sample processing
    procedure, 198–99
  clines, 128
  DNA ancestry testing methods, 207–9
  Greek aDNA studies, 180
  overview of, 33*b*
Sirugo, Giorgio, 126–27
skin color
  biological characteristics distinguishing
    non-biological races, 58–60
  differences in degree rather than
    kind, 59–60
  environmental clines, 157–58
  evolution of, 61*b*
  formation of, 58*b*
  selective sweeps, 140*b*
slavery, 17
  admixture, 169–70
  Jefferson/Hemings case, 95–97
  justifying, 53–54, 155
  racist exploitation, 54
SNPs. *See* single nucleotide
  polymorphisms
social race, defined, 4. *See also* race
Sognnaes, Reidar Fauske, 94
Sommer, Marianne, 169–70
sperm donors, 6–7
stability. *See* fixity and stability
Stephanopoulos, George, 201–2
Stoneking, Mark, 104
Stormfront.org, 21
strontium isotope analysis, 189–91, 193
STRUCTURE method, 130–33,
  131*f*, 135–36
supra-orbital ridge (brow ridge) and skull
  comparisons, 52–53, 53*f*
Sykes, Bryan, 105–6, 106n.8
Systema naturae (Linnaeus), 50

Tallbear, Kim, 27
Tatiana (grand duchess), 97–98
Tattersall, Ian, 147–48, 164–65

Thomas Jefferson Memorial Foundation,
  Inc., 97
Tishkoff, Sarah A., 126–27
Transatlantic Slave Trade. *See* slavery
Triantaphyllidis, Costas, 177–78
Troublesome Inheritance (Wade), 137
Trump, Donald, 26–27
23andMe
  number of profiles in database, xi
  promoting DNA testing, xi–xii
  self-reported ancestry vs. test
    results, 9–10
  testing for both ancestry and health, 6
  unequal representation in
    database, 10–11
  varied results among companies, 11–12,
    13*t*, 14–15
typological thinking, 50–51, 54–55

United Nations Educational, Scientific,
  and Cultural Organization
  (UNESCO), 64
University College, Dublin, 76–77
University of Oxford, 106n.8
US Census
  ethnicity, 10
  population estimates by race, 10–11
  social construction of race, 62–63

Veeramah, Krishna, 193
Victoria (queen), 97–98, 100

Wade, Nicholas, 137–42, 143
Waj (*The DNA Journey* participant), 102–3
Washburn, Sherwood, 54–55
Washington, George, 95–96
Wayles, John, 95–96
White people
  effects of test results on self-perceptions
    and identity, 17, 18
  essentialism, 51–52
  "one drop rule," 62
  perceptions of and attitudes toward
    testing, 17
  social construction of category, 62
  social construction of race, 63

US Census descriptions, 63
variation in designation among
studies, 154–55
White French Canadians and claims of
Indigenous identity, 21–22
White supremacy and nationalism,
7–9, 21
Who We Are and How We Got Here
(Reich), 149
Williams, Scott M., 126–27
Wilson, Allan, 104
Winfrey, Oprah, 54
Woodson, Thomas, 96

Xenia Cheremeteff-Sfiri
(countess), 98–100

Y-Adam, 106–7

Yale School of Medicine, 12–14
YDNA
autosomal DNA vs., 87*b*, 117,
120–21
coalescence time, 106–7
"deep" ancestry, 117–21
haplogroups, 118–20
haplotypes, 118
inheritance of, 86, 87*b*, 89*f*
Jefferson/Hemings case, 96, 97
limitations of focus on, 120–21
mtDNA vs., 88*b*, 196
representation of ancestry in, 90
Romanov case, 100
Y-Adam, 106–7
Yuval-Davis, Nira, 25

Zeruvabel, Eviatar, 75